AIRCRAF
WEIGHT AND BALANCE

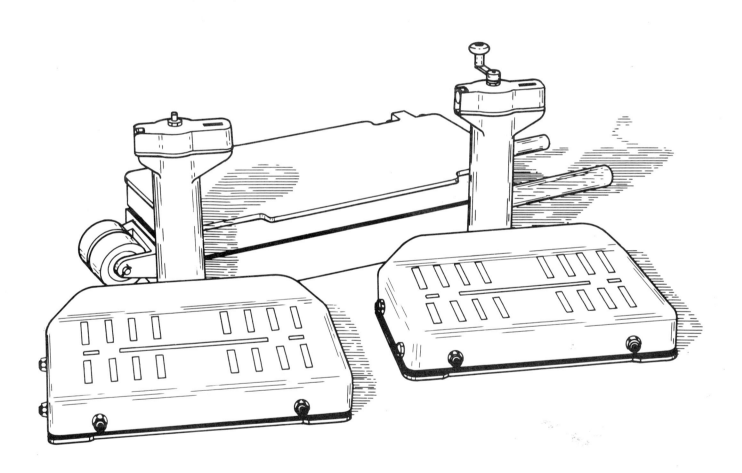

JEPPESEN®
Sanderson Training Products

Table of Contents

PREFACE

This book on *Aircraft Weight and Balance* is one of a series of specialized training manuals prepared for aviation maintenance personnel.

This series is part of a programmed learning course developed and produced by Jeppesen Sanderson, Inc., one of the largest suppliers of aviation maintenance training materials in the world. This program is part of a continuing effort to improve the quality of education for aviation mechanics throughout the world.

The purpose of each training series is to provide basic information on the operation and principles of the various aircraft systems and their components.

Specific information on detailed procedures should be obtained from the manufacturer through the appropriate maintenance manuals, and followed in detail for the best results.

This particular manual on *Aircraft Weight and Balance* includes a series of carefully prepared questions and answers to emphasize key elements of the study, and to encourage you to continually test yourself for accuracy and retention as you use this book.

Some of the words may be new to you. They are defined in the Glossary found at the back of the book.

For product, service, or sales information call 1-800-621-JEPP, 303-799-9090, or FAX 303-784-4153. If you have comments, questions, or need explanations about any Maintenance Training System, we are prepared to offer assistance at any time. If your dealer does not have a Jeppesen catalog, please request one and we will promptly send it to you. Just call the above telephone number, or write:

Marketing Manager, Training Products
Jeppesen Sanderson, Inc.
55 Inverness Drive East
Englewood, CO 80112-5498

Please direct inquiries from Europe, Africa, and the Middle East to:

Jeppesen & Co., GmbH
P. O. Box 70-05-51
Walter-Kolb-Strasse 13
60594 Frankfurt
GERMANY
Tel: 011-49-69-961240
Fax: 011-49-69-96124898

Introduction

From the earliest days of aviation it was learned that *weight* was one of the determining factors in the ability of an aircraft to fly. The builders of these early aircraft quickly adapted to the use of such materials as wood, dope and fabric to obtain the strength to weight ratios that would allow flight.

However, during this early development period little thought was given to *balance*. This oversight resulted in failure and often catastrophic events leading to the death of several would-be designers, builders, and pilots.

By World War I it had become common knowledge among designers that in order to obtain a good aircraft it must be light and maneuverable. The lightness was obtained by the materials used in construction and the maneuverability could be increased by placing the center of gravity directly in line with the center of lift.

As the aircraft developed, and more uses evolved, it became more complex, carried more than one man, and traveled greater distances, causing more problems to be faced by the aircraft builder. One of these problems was *stability*. By the addition of passengers, navigational aids, fuel, and baggage, it became quite evident that the aircraft could loose stability simply by moving these items into different locations within the aircraft. Sometimes attitude changes also brought about a complete loss of controlability. This brought about exotic devices such as a "Flat Spin Parachute". This was placed in the tail in order to gain control when such instances occured.

By the late 1930's it became quite obvious that the aircraft must be designed in such a manner that the center of gravity must be placed slightly ahead of the center of lift. Although this reduced the maneuverability, it resulted in a stable aircraft with less chance of loss of control.

During World War II, more reliable, larger and faster aircraft were needed more than ever before. Aircraft designers met these demands with the use of aluminum structures, more engines, and cleaner aerodynamic designs. As a result, a more complex aircraft was developed. These aircraft were capable of lifting thousands of pounds, going to altitudes requiring pressurization systems, and traveling at speeds in excess of three hundred miles per hour.

By the end of this period the aircraft became an accepted mode of transportation by the public with further demands for more speed, comfort, and conveniences. This brought about the use of the turbine engine, the swept wing and speeds approaching "Mach 1". These new aircraft concepts posed even more weight and balance problems to the designer. Fuel had to be used in a proper sequence, food facilities had to be provided for the passengers, a wide range of fluids were now used in the various systems, and the center of lift would actually move as we approached the speed of sound.

Todays designers must supply an aircraft to meet every facet of transportation and pleasure. These include small training aircraft, helicopters capable of takeoff and landing in small areas, transports carrying loads in excess of two hundred thousand pounds, and aircraft able to carry passengers to their destination at speeds greater than the speed of sound. Although each of these aircraft has brought about its own unique features regarding weight and balance, there are common rules that govern safe flight.

As we will see in this book, all of these aircraft will be adversely affected by improper loading. This may be due to an excess of weight or an improper placement of the weight in different locations within the aircraft.

Effects of an overloaded aircraft are as follows:

1. More runway will be needed.
2. A lower climb angle and higher speed will be required.
3. Structural safety factors are reduced.
4. Stalling speeds are increased.
5. More engine power is required.

Effects on an aircraft with too much weight forward are as follows:

1. Tendency to dive forward occurs.
2. Stability is decreased.
3. Adverse spin characteristics occur.
4. More engine power is required.

Effects on an aircraft with too much weight aft are as follows:

1. Flying speed is decreased.
2. Stall characteristics occur.
3. Stability is decreased.
4. Adverse spin characteristics occur.
5. More engine power is required.

Any of the conditions mentioned above could result in the loss of the aircraft and loss of life. For these reasons, it is very important that the aircraft technician and the pilot have a thorough understanding of weight and balance of the modern aircraft.

CHAPTER I

Theory of Weight and Balance

All aircraft have design limitations. One of these limitations is *weight*. The factors that must be included in determining the weight limitations are the structure, the ability to lift the aircraft, and the maneuvers that the aircraft is allowed to perform. For these reasons a *maximum weight* must be established for each aircraft design.

A. Maximum Weight

Maximum weight is the *total* weight of the aircraft, equipment, passengers, baggage and fuel. Often we will find that an aircraft has two or more maximum weights. This will be determined by the category in which the aircraft is operated. For example, the maximum weight for a particular aircraft is 2550 pounds in the normal category and 2000 pounds in the utility category. This difference in weight is entirely due to the maneuvers the aircraft is allowed to perform in the two different categories.

Large transport type aircraft often have three maximum weights: a taxi or ramp weight, a takeoff weight, and a landing weight. These three weights are necessary to maintain lift and structural integrity during its different phases of operation rather than maneuverability limitations.

Some of these aircraft carry additional fuel for taxi purposes because of our congested large airports and large aircraft. This could be as great as 4000 pounds above the takeoff weight. This additional weight is referred to as taxi or ramp weight.

Takeoff weight is the greatest amount of weight that the aircraft is safely capable of lifting from the ground. In the transport aircraft category this may also be limited by the length of the runway and runway conditions. Atmospheric conditions such as temperature and barometric pressure will also affect the power output of the engines.

Maximum landing weight is the greatest amount of weight with which the aircraft can land safely.

We will find such weight limitations in the transport aircraft category. The greatest weight change in these aircraft will be fuel consumption. In aircraft with a greater takeoff weight than landing weight a fuel dump system must be utilized in the aircraft design in order to ensure a safe landing. For our purposes in this book we will refer to the maximum weight as the maximum authorized weight of the aircraft and its contents.

B. Empty Weight

The Empty Weight of an aircraft is the weight of the airframe engine and all equipment that has a fixed location or is installed in the aircraft. It does not include passengers, baggage and fuel. However, fuel trapped in the system after draining is included in the Empty Weight.

Oil may or may not be included in the Empty Weight of the aircraft. For many years, oil has not been part of the empty weight except for residual oil or undrainable oil. Today, due to a change in the FAR 23 aircraft are being manufactured which include full oil as a part of the Empty Weight.

To determine whether a specific aircraft's Empty Weight includes engine oil the Aircraft Data Sheet must be checked.

Other operating fluids, such as the hydraulic fluid, are included in the Empty Weight.

In most cases, the manufacturer only weighs every tenth aircraft in order to establish the empty weight of a particular type of aircraft. This is done prior to adding the optional equipment. Then, mathematically, the optional equipment with a fixed location is added to the empty weight. This weight, without optional equipment, is referred to as the *basic weight* and should not be confused with the empty weight.

Another term that is sometimes considered with empty weight or basic weight is the *operating weight*. Operating weight is used on transport types of aircraft where certain items are always

1

carried on the aircraft. The crew, galley water, survival gear, fuel. oil, and de-icing fluids. water injection etc. that are not part of the empty weight fall into this category. Operating weight is used to facilitate faster loading calculations.

C. Useful Load

The *useful load* is the empty weight of the aircraft subtracted from the maximum weight of the aircraft. This includes the oil, fuel, crew, passengers, baggage, and cargo. The useful load and weight distribution is often determined by the category under which the aircraft is operated. For example, an aircraft may have a useful load of 3000 pounds in the normal category and 2500 pounds in the utility category.

Often the useful load is changed to meet the needs of the specific flight. For example, on a short flight, a pilot may choose less fuel in order that more cargo may be carried and still remain within the maximum weight. On a longer flight the pilot may choose more fuel and less cargo.

All useful loads must also be kept within structural and balance limits. For instance, most aircraft are built to carry a certain floor loading and a limited weight in certain areas. If too much weight is placed in the nose or the tail of the aircraft adverse flight characteristics will occur.

For weight and balance purposes the FAA has assigned specific weights to the crew and passengers, gasoline, oil and turbine fuel. These are as follows:

Crew and Passengers	170 pounds per person
Gasoline	6 pounds per U.S. Gallon
Oil	7.5 pounds per U.S. Gallon
Turbine Fuel	6.7 pounds per U.S. Gallon

D. Datum

The *datum* is an imaginary line on a vertical plane from which all horizontal measurements on the aircraft are taken for weight and balance purposes. These measurements are taken with the aircraft in a level flight position. From this datum we can determine the distances for the location of such items on the manufacturer's equipment lists such as seats and special equipment. It can also be used when new equipment is to be added or old equipment is to be removed from the aircraft.

The actual location of the datum for a particular type of aircraft can be any point selected by the

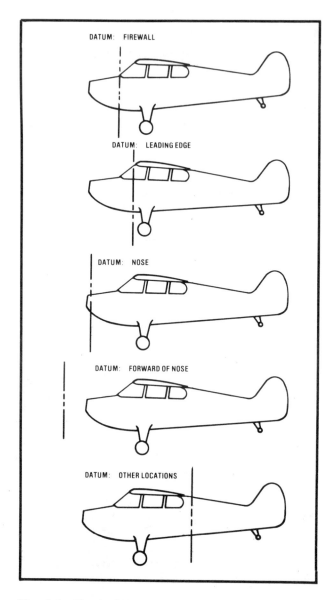

Fig. 1-1 Typical Datum Locations.

manufacturer. Common places are the leading edge of the wing, the firewall and the nose. There is a tendancy today for the manufacturers to place the datum forward of the nose of the aircraft. On a few of the older aircraft no datum was selected. In such cases any point may be used as long as all measurements are calculated from that same point. (Fig. 1-1)

E. Arm

The *arm* is the horizontal distance that an item is located from the datum. This distance is always given in inches. If the particular item is located forward of the datum it is shown with a negative (−) sign. If the item is located aft of the datum it is shown with a positive (+) sign.

2

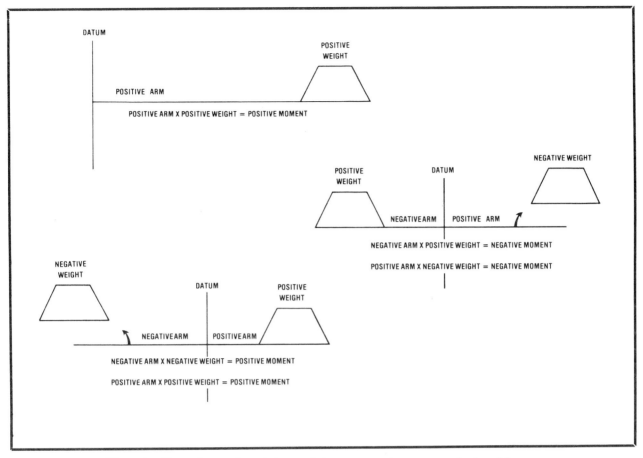

Fig. 1-2   Negative and positive weight, arm, and moment relationships.

Many items used in our weight and balance computations will have arm distances furnished by the manufacturer or will be found in the FAA Specifications. For example, these will be shown as (+25) or a (−50) meaning 25 inches aft of the datum and 50 inches forward of the datum respectively. If the distance for an item to be installed is not given, an actual measurement must be taken.

F.   Moment

Moment is the product of the weight multiplied by the arm. This measurement of force will be in inch pounds. The longer the distance from the datum the larger the moment will become. For example, 5 pounds placed 25 inches from the datum will have a moment of 125 inch pounds. Five pounds placed 100 inches from the datum will have a moment of 500 inch pounds.

Moment may be either negative or positive. This will be determined by whether the weight is added or removed and whether the arm is negative or positive.

When weight is added it will be positive. When weight is removed is will be negative. The sign of the arm as we have previously discussed will depend on its relation to the datum: negative when forward of the datum and positive when aft of the datum. The moment will be positive if the arm and weight are both positive. It will also be positive if the arm and the weight are both negative. The moment will be negative if the arm is negative and the weight is positive or if the arm is positive and the weight is negative. The easy way to remember this is: "like signs result in positive numbers and unlike signs will result in negative numbers." Moment is expressed as a number. For example: moment +500 or ¹500. (Fig. 1-2)

There are some distinct advantages in placing the datum ahead of, or at the nose of the aircraft because it will result in all of the arm measurements becoming positive numbers. This means that the only negative moments will be obtained by removing equipment.

3

However, there is one disadvantage to this datum location. If actual measurements for placing items in an aircraft must be made, it is quite difficult to take these measurements without a datum point on the aircraft.

G. Center of Gravity

The *center of gravity* (commonly abbreviated CG) is the point at which the nose heavy moments and the tail heavy moments have equal magnitude. This would be the point at which the aircraft could be suspended without having any tendency to become nose or tail heavy (Fig. 1-3).

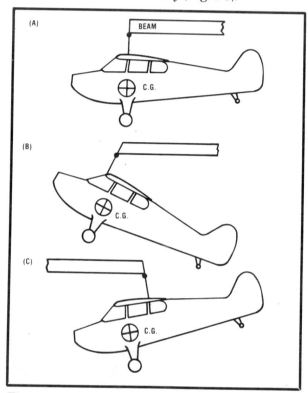

Fig. 1-3 A — Aircraft is suspended from its CG. B — Aircraft is tailheavy. C — Aircraft is noseheavy.

We could possibly suspend a light aircraft from various points to locate the CG but this would be highly impractical and also impossible with a large aircraft. So, for practicality, it must be done mathematically. The formula for obtaining the center of gravity is the total moment divided by the total weight which may be abbreviated CG = $\dfrac{TM}{TW}$

Other formulas may be used to obtain the center of gravity of an aircraft, as we will discuss later, but these variations will always utilize the total moment divided by the total weight formula.

This formula may be used to find the CG of any object. For the sake of simplicity we will start with weight and arms as in a teeter-totter arrangement and the CG as the point of balance (Fig. 1-4).

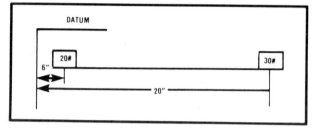

Fig. 1-4 Graphic description of Problem 1.

PROBLEM 1

As we mentioned previously, the datum could be located at any point. In this problem (see Fig. 1-4) our distance is 6 inches forward of the 20 pound weight and 20 inches forward of the 30 pound weight. Therefore, our arms are 6″ and 20″ respectively. The formula for determining moment is: weight × arm.

Both weights are positive and both arms are positive, so our moment will also be positive. The moments will be 6 × 20 which equals 120 and 20 × 30 which equals 600. The total moment will be 120 + 600 which will equal 720. The total weight is 20# + 30# which equals 50 pounds. By dividing total moment by total weight we will place the CG at 14.4″ from the datum. (Numbers should be rounded off to the nearest tenth.)

PROBLEM 2

Using another teeter, a different datum and weights, we will work another problem (Fig. 1-5).

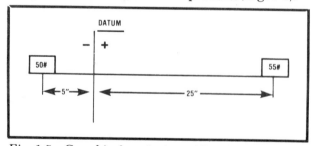

Fig. 1-5 Graphic description of Problem 2.

Our datum is now 5 inches aft of the 50 pound weight and 25 inches forward of the 55 pound weight. Using the *arm × weight formula* to determine moment, the forward weight will result in a −250 moment because the sign of the arm is

4

negative. The moment of the aft weight will be +1375 moment. Since we have a negative and positive moment, the total moment can be found by subtracting the negative from the positive leaving 1125 positive moment. This divided by the weight of 105 will place the CG at the 10.7"aft of the datum.

PROBLEM 3

By moving our datum to a new location of 25 inches aft of the forward weight of 50 pounds we have a new problem (Fig. 1-6).

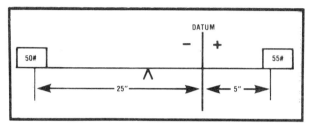

Fig. 1-6   Graphic description of Problem 3.

The weight has remained the same but the arms have changed because of the change in the datum location. Therefore, the moment for the 50 pound weight will be a negative number. The weight is a positive number. The moment must be negative or +50 × −25 = −1250.

The 55 pound weight is aft of the datum and the weight is positive, so the moment is positive or 5 × 55 = 275. The two moments must be sub-

tracted: −1250 − +275 = −975 moment. Total moment divided by total weight will leave the CG at −9.3 inches ahead of the datum.

Returning to the second problem we will find the distance from the datum to the CG was 10.7 inches. The distance from the 50 pound weight to the datum was 5 inches making a total of 15.7 inches. The datum in the third problem was 5 inches from the 55 pound weight. The CG was 9.3 inches ahead of the datum or a total of 14.3 inches from the weight. This proves that the location of the datum makes no difference (Fig. 1-7).

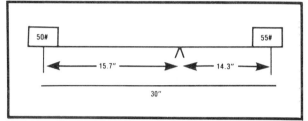

Fig. 1-7   Proof that the datum location makes no difference.

H.   Center of Gravity Range

It is quite easy to see that since we have found the center of gravity, each time an aircraft is built or each time the aircraft is loaded, we cannot locate the CG in exactly the same place.

Because of these variations a *CG range* must be established. The limits of the CG range are established by the manufacturer for a most forward

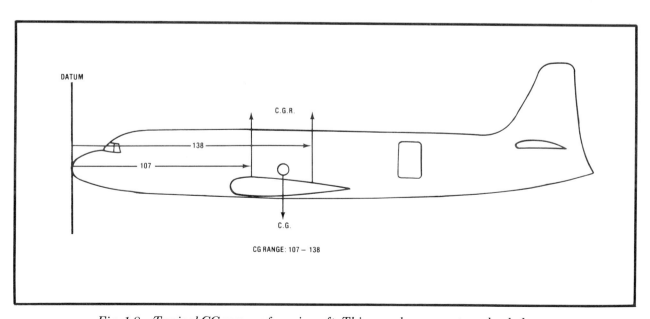

Fig. 1-8   Typical CG range of an aircraft. This may be an empty or loaded range.

and rearward CG that will allow a safe flight of the aircraft (Fig. 1-8).

Often an aircraft will have two CG ranges which is determined by its operation in either the normal or utility category. The utility category is generally less because of the maneuvers which are allowed in this category.

There are two CG ranges that have been established. These are the *empty CG range* (ECGR) and the *operating CG range* (OCGR).

The empty CG range is the established limits that the empty CG falls within. It is not possible to load the aircraft in such a manner as to adversely affect the flight characteristics if it is within these limits. Unfortunately most aircraft do not have an empty CG range.

The operating CG range is the established CG limits forward and aft in the loaded aircraft. These limits must be strictly observed by the pilot at all times during flight and he must distribute his load accordingly. This information may be made available to him by placards, loading charts, or load adjusters which will be discussed in detail later in this book.

The center of gravity range is expressed in a percentage of *mean aerodynamic chord* (MAC).

The MAC is the mean (average) chord of the wing. This measurement is actually the determining factor used by the manufacturer to select the CG location on aircraft. For simplicity purposes most light aircraft manufacturers express the CG range in inches from the datum.

Today, on transport aircraft, irregularly shaped wings are used almost exclusively. In most cases a taper, crescent, swept back or swept forward wing or a combination of these types are used. For this reason it is advantageous to express the CG range in MAC percentages rather than in inches from the datum (Fig. 1-9).

I. Weighing Points

In order to find the center of gravity, *weighing points* must be selected. For placing the scales, the most commonly used points are the wheels or the jack points of the aircraft. Either of these places are designed to support the weight of the aircraft. Usually on light aircraft the landing gear is used. This would be difficult on large aircraft so the jack pads must be used (Fig. 1-10).

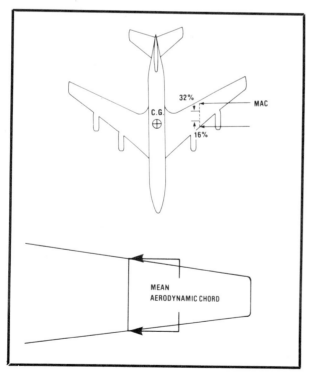

Fig. 1-9 Mean Aerodynamic Chord of a large aircraft.

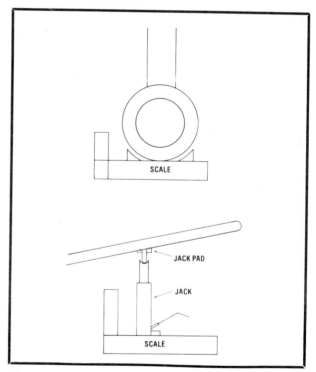

Fig. 1-10 The aircraft may be weighed from the main gear or the wheels.

J. Aircraft Leveling Means

In order to find the CG of an aircraft it must be level. For this purpose a leveling means is provided by the aircraft manufacturer. It may be nothing more than a door sill, or it could be two lugs built into the fuselage in order to accomodate a spirit level. On some aircraft it consists of a point to suspend a plumb bob over a scale (Fig. 1-11).

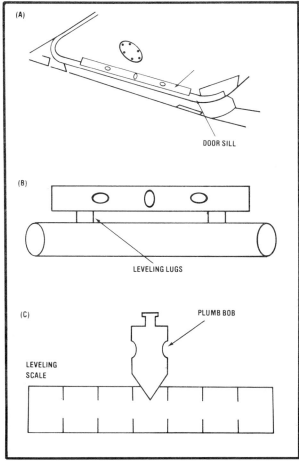

Fig. 1-11 Typical leveling means. A — Door sill. B — Leveling lugs. C — Leveling scale.

K. Minimum Fuel

It is sometimes necessary to check the most forward and aft center of gravity range limits in a loaded type of condition. Since fuel is used during flight the aircraft may have CG problems when the fuel supply becomes low. For this reason we use a measurement referred to as *minimum fuel* which is used for calculating adverse forward and rearward CG conditions. It is also sometimes referred to as *maximum except takeoff fuel* (METO fuel).

This measurement on older aircraft, built under CAR 3, is 1/12 of a gallon per maximum continuous horsepower of the engine. The simplest way to derive this amount for a gasoline engine is to divide METO power by 2 which will give you the amount of fuel in pounds. Since gasoline weighs 6 pounds per gallon, METO power divided by 6 will give the gallons necessary for METO fuel. For example: an aircraft with 220 hp maximum takeoff would require 110 pounds of gasoline (220 divided by 2 equals 110) and would require 18.3 gallons. (110 divided by 6 equals 18.3). These formulas should not be used for the turbine powered aircraft because the aircraft manufacturer designates minimum fuel for the turbine aircraft.

On new aircraft built under FAR 23, a different rule is used for the calculation of minimum fuel. It is the fuel necessary for 1/2 hour of flight at rated maximum continuous power.

QUESTIONS:

1. Takeoff weight and landing weight are always the same weight.

    A. True.

    B. False.

2. The maximum weight at which an aircraft may operate is the same regardless of the category in which the aircraft is operated.

    A. True.

    B. False.

3. The empty weight of the aircraft includes fuel and oil.

    A. True.

    B. False.

4. Hydraulic fluid is considered part of the empty weight.

    A. True.

    B. False.

5. The term operating weight is synonomous with useful load.

    A. True.

    B. False.

6. As long as the useful load is not exceeded, it makes no difference where it is placed.

    A. True.

    B. False.

7. Turbine fuel is heavier than gasoline.

    A. True.

    B. False.

8. The location of the datum is always at the leading edge of the wing.

    A. True.

    B. False.

9. All measurements forward of the datum are negative.

    A. True.

    B. False.

10. The arm is the horizontal distance an item is from the datum.

    A. True.

    B. False.

11. Moment is weight times the arm.

    A. True.

    B. False.

12. Equipment removed from the aircraft always carries a + sign.

    A. True.

    B. False.

13. Moment may be positive or negative.

    A. True.

    B. False.

14. If the datum is located at the nose of the aircraft all moments are negative.

    A. True.

    B. False.

15. The point at which the negative and positive moments are equal is referred to as the center of moments.

    A. True.

    B. False.

16. The center of gravity of an aircraft will remain the same throughout the flight.

    A. True.

    B. False.

17. Some CG ranges are measured in % of MAC.

    A. True.

    B. False.

18. All aircraft are weighed by placing scales under the wheels.

    A. True.

    B. False.

19. The door sill is the standard point to level an aircraft.

    A.   True.

    B.   False.

20. The formula for minimum fuel weight is METO divided by 2.

    A.   True.

    B.   False.

CHAPTER II

Data Investigation

In order to weigh and to determine the center of gravity of an aircraft, one must be able to obtain and interpret information about the specific aircraft. It would be quite simple if all of this information could be obtained from one source, but unfortunately, this is not possible.

As the aircraft developed, so did the pertinent data for each aircraft. Today, our two primary sources of information are the Federal Aviation Agency and the aircraft manufacturer.

A. FAA Materials

The FAA is responsible for determining that each designed aircraft is built in accordance with the *Federal Air Regulation* under which that particular aircraft was designed and certified. Once the aircraft is certified and production begins, the FAA will issue an *aircraft data sheet*. This will contain pertinent information about this particular aircraft built under this type of certification. As new models are built of the same aircraft this data sheet is revised by the FAA.

1. Aircraft data sheets

The data sheet will include such information as: location of the datum; maximum weight of the aircraft; empty center of gravity range (if any); loaded center of gravity range for the particular category under which it is to be operated; leveling means; maximum baggage and location; seat location; engine horsepower; fuel and oil capacity. All of the above information will be necessary in the determination of the *weight and balance* of a particular aircraft. Any additional information for our calculations is furnished by the manufacturer. This includes the approved equipment, optional equipment and their weight and locations (Fig. 2-1).

It should be noted in our illustration, that the same data sheet covers several different models

of the 172. The data sheet is broken down into particular models with the details pertinent to the particular model in the first part of the data sheet along with the information on the engine, fuel and propeller. Usually this information will not be used for the weight and balance computations except when the engine information is needed for METO calculations.

The next area of the data sheet includes the CG ranges, maximum weight, number of seats, baggage, fuel, and oil. These will be necessary for the weight and balance calculations. In reviewing these figures it can be observed that the CG range varies with the category in which the aircraft is operated and the weight of the aircraft. There will also be a difference if the aircraft is operated with floats.

On this particular aircraft, no empty CG range is given. This is quite typical of our more modern aircraft. On this particular aircraft, loading charts are used by the pilot to determine his loaded CG. On other aircraft, if the empty CG falls within limits, no other computations were necessary for flight as long as normal loading was utilized. A few are still built in this manner today.

The maximum weight on this data sheet varies between the normal and the utility. There is also a variation for land and float operation. On transport type aircraft there will be a maximum takeoff weight and a maximum landing weight. There may also be a *taxi weight* listed on this type of aircraft.

In the number of seats section, the seating arrangement is given with reference to station location which is always given with a plus or a minus sign depending on its sign from the datum. Optional seating will often be given for different configurations of the same aircraft. This optional seating may also be given in the form of a *note* which will be found towards the back of the data sheet.

```
                                                          3A12
                                                   Revision 43
                                                      CESSNA
                        172                              172I
                        172A                             172K
                        172B                             172L
                        172C                             172M
                        172D                             172N
                        172E
                        172F (USAF T-41A)
                        172G
                        172H (USAF T-41A)
                                              October 15, 1977
```

VII — Model 172M, 4 PCL — SM (Normal Category), 2 PCLM (Utility Category), approved May 12, 1972

*Engine	Lycoming 0 — 320 — E2D
*Fuel	80/87 minimum grade aviation gasoline
*Engine limits	For all operations, 2700 rpm. (150 hp.)

Propeller and
  propeller limits

1. Propeller
    (a)   McCauley 1C160/CTM7553
        Static rpm at maximum permissible throttle setting:
           Not over 2370, not under 2270
           No additional tolerance permitted (See Note 3)
        Diameter: Not over 75 in., not under 74 in.
    (b)   Spinner: Dwg. 0550320

2. Propeller
    (a)   McCauley 1C160/DTM7553
        Static rpm at maximum permissible throttle setting:
           Not over 2370, not under 2270
           No additional tolerance permitted (See Note 3)
        Diameter: Not over 75 in., not under 74 in.
    (b)   Spinner, Dwg. 0550320

3. Propeller (seaplane only)
    (a)   McCauley 1A175/ATM8042
        Static rpm at maximum permissible throttle setting:
           Not over 2545, not under 2445
           No additional tolerance permitted (See Note 3)
        Diameter: Not over 80 in., not under 78.4 in.
    (b)   Spinner, Dwg. 0550320

4. Propeller (seaplane only)
    (a)   McCauley 1A175/ETM8042
        Static rpm at maximum permissible throttle setting:
           Not over 2545, not under 2445
           No additional tolerance permitted (See Note 3)
         Diameter: Not over 80 in., not under 78.4 in.
    (b)   Spinner, Dwg. 0550320

*Airspeed limits (TIAS) — 17256493, 17260759 through 17265684

Maneuvering	112 mph.	(97 knots)
Maximum structural cruising	145 mph.	(126 knots)
Never exceed	182 mph.	(158 knots)
Flaps extended	100 mph.	(87 knots)

*Airspeed limits (IAS)
(See Note 4 on use of IAS) — 17265685 through 17267584

Maneuvering	97 knots
Maximum structural cruising	128 knots
Never exceed	160 knots
Flaps extended	85 knots

Fig. 2-1   Sample data sheet.

11

C.G. range	Landplane:	
	Normal category	(+38.5) to (+47.3) at 2300 lb. (+35.0) to (+47.3) at 1950 lb. or less
	Utility category	(+35.5) to (+40.5) at 2000 lb. (+35.0) to (+40.5) at 1950 lb. or less

Seaplane: (EDO 89-2000 or 89A2000 floats)
   Normal category   (+39.8) to (+45.5) at 2220 lb.
                                          (+36.4) to (+45.5) at 1825 lb. or less
Straight line variation between points given.

Empty wt. C.G. range	None

*Maximum weight	Normal category:	2300 lb. (landplane); 2220 lb. (seaplane)
	Utility category:	2000 lb. (landplane)

No. of seats 4 (2 at +34 to +46, 2 at +73) (Occupant on child's optional jump seat at +96)

Maximum baggage 120 lb. at +95

Fuel capacity 42 gal. total, 38 gal. usable (two 21 gal. tanks in wings at +48). See Note 1 for data on unusable fuel.

Oil capacity 2 gal. (−14.0), 1-½ gal. usable. See Note 1 for data on undrainable oil.

Data Pertinent to All Models:

Datum Front face of firewall (28000 through 47746)
Lower front face of firewall (17247747 and on)

Leveling means Upper door sill

Note 1.   Current weight and balance report including list of equipment included in certificated empty weight and loading instructions when necessary must be provided for each aircraft at the time of original certification.

Serial Nos. 28000 through 29999, 36000 through 36999 and 46001 through 47746, 17247747 through 17265684
The certificated empty weight and corresponding center of gravity location must include unusable fuel fo 30 lb. at (+46) for Models 172 and 172A, or 18 lb. at (+46) for Models 172B through 172H, or 24 lb. at (+46) for Models 172I through 172M (17265684) and undrainable oil of (0) lb. at (−20) for 172 through 172H and (0) lb. at (−14) for 172I through 172M (17265684).

Serial No., 17261578, 17261445, 17265685 and on
The certificated empty weight and corresponding center of gravity location must include unusable fuel of 24 lb. at (+46) through 172M (17267584) or 18 lb. at (+46) 17267585 and on and full oil of 11.3 lb. at (−14).

Fig. 2-1   (continued)

The maximum baggage is given in pounds along with the station location which always includes the sign of the arm. For example: (+95).

The fuel capacity is given in gallons with the arm and its sign. This area of the data sheet may also include undrainable fuel and unusable fuel and optional fuel tanks. On this particular data sheet they refer to *note 1* for further information. As previously mentioned, this will be contained in the latter part of the data sheet because the same information applies to several models of the 172. *Remember, unusable fuel is part of the empty weight and is given in the data sheets in pounds with the arm locations.*

The oil capacity is given in quantity with the arm location and the usable oil quantity. On the particular data sheet the undrainable oil is included in note 1. The undrainable oil is part of the empty weight of the aircraft. *Unlike the fuel, however, unusable oil is not part of the empty weight.*

2. Aircraft specification sheets

The FAA issued *aircraft specification sheets* on all aircraft that were type certificated prior to 1963. The aircraft specification sheet differs from the aircraft data sheet because the approved required and optional equipment was listed with the pertinent information of the aircraft. This was a great advantage to the maintenance personnel because only one source of information was needed. However, it was quite difficult to keep this information updated and it required costly revisions (Fig. 2-2).

This information is basically the same as is contained in the data sheet. However, it also includes the *required* and *approved* optional equipment. This information will be of great value if the equipment list must be replaced, or if new items of equipment are to be added or removed from the existing equipment. This type of change to the existing weight and balance is generally done on paper rather than actually reweighing the aircraft.

The first variation from the data sheet that is to be noted is a *list of required equipment*. This list of numbers corresponds with the approved equipment list contained in the specification sheet, and means that the items have *alternate* equipment in lieu of a required item.

Preceeding the equipment in the specification sheet is a statement: "A plus (+) or minus (−) sign preceeding the weight of an item indicates a net weight change when that item is installed." The items that are referred to are *alternate items*, and if they are used to replace previously installed equipment, it is the difference in weight between the old item and the new item.

As in the data sheets, the specifications cover more than one model of the same aircraft. This is reflected in the equipment section by an *unbroken line* if the item is not applicable to the particular model, and the arm length, if it does apply. For example: (+3).

The actual weight change is required on a few items, such as a fuel injection installation.

3. Aircraft listings

The FAA issues *aircraft listings* for aircraft that are no longer in production and number less than fifty in the *Aircraft Registry*. These contain basically the same information as the specification sheet. (Refer to Figure 2-3.) They are similar with two exceptions which are concerning the equipment listed in certain classes. Class I equipment should be considered as *required equipment* while Classes II and III are considered *optional equipment*.

The other exception is an asterisk (*), which is used to designate net changes in weight rather than the plus or minus signs used in the aircraft specifications.

B. Manufacturers' Materials

Since the inception of the Aircraft Specification Data Sheet, the manufacturer of the aircraft has been responsible for furnishing the additional information required for weight and balance purposes. To date, there is no standardization among the manufacturers in this information and it is rather doubtful that it will occur in the near future.

The *weight and balance data* and an *equipment list* are required with each aircraft. The weight and balance data is necessary for pilot computation of loading the aircraft. The equipment list is necessary so that an accurate weight and balance may be kept with changes made to the aircraft by the addition or removal of equipment.

13

```
1A15
Revision 28
PIPER
PA — 24
PA — 24 — 250
PA — 24 — 260
PA — 24 — 400
May 1, 1974
```

<u>Required Equipment.</u> In addition to the pertinent required basic equipment specified in CAR 3, the following items of equipment must be installed:

3(a) and (b), 103, 106, 107 (a) or (b), 201 (a), 202 (b), 205, 206 (a), 301 (a) or (b), 302 (a), 401 (c), or (f) or (n) or (s) or (v), 603, 617 (a), 401 (bz).

When Item 109(e) is installed, substitute Items 113 and 114 (a) or 114(b) in place of Items 106 and 107 (a) or 107 (b) respectively.

<u>Miscellaneous Equipment</u> (Not Listed Above)			PA-24	PA-24-250	PA-24-400	PA-24-260
601.	Heated Pitot Head-Kollsman, Model NO. 372D-01, 12 volt.	+ 1 lb.	(+99)	(+99)	(+99)	-----
602.	Heated Pitot Tube Assembly, PAC Dwg. 21301	+ 1 lb.	(+99)	(+99)	(+99)	(+99)
603.	Stall warning indicator installation installed in accordance with PAC Dwg. 21754.	Negl.Wt.	-----	Required	Required	Required
604.	Stabilizer guard installation in accordance with PAC Dwg. 22004.	+ 1 lb.	(+255)	(+255)	-----	(+255)
605.	Manifold pressure gauge. NOTE 2(k) placard required on Model PA-24-400.	+ 1 lb.	-----	-----	Required	See Item 9(a) or 9(c)
606. (a)	Piper Radio Coupler Per PAC Dwg. 24997 or 23619.	+ .5 lb.	-----	-----	(+66)	(+66)
(b)	Piper Radio Couler Per PAC Dwg. 25678, NOTE 2(p) placard required.	+ .5 lb.	-----	-----	-----	(+66)
(c)	Piper Radio coupler per PAC Dwg. 26825 NOTE 2(r) placard required.	+ .5 lb.	-----	-----	-----	(+66)
607. (a)	Exhaust Gas Temperature Indicator per STC SA788WE, installed per Pac Drawing 25287.	+ 1 lb.	-----	-----	(+55)	(+55)
(b)	Exhaust Gas Temperature Indicator per STC SA522SW, installed per PAC Drawing 25667.	+ 1 lb.	-----	-----	-----	(+55)
(c)	Exhaust Gas Temperature Indicator per STC SA1315WE, installed per PAC Drawing 26333.	2 lb.	-----	-----	-----	(54)
(d)	Exhaust Gas Temperature Indicator per STC SA1315WE, installed per PAC Dwg. 26851.	2 lb.	-----	-----	-----	(+54)
608. (a)	Alternate instrument static air source installed per PAC Dwg. 25301 and 26723. NOTE 2(1) placards required.	Neg.Wt.	-----	-----	Elig.	Elig.
609.	Optional engines (Lycoming)					
(a)	0-540-A1A5 (Item 401(f) required)	Neg.Wt.Ch.	-----	Elig.	-----	-----
(b)	0-540-A1B5 (Item 401(f) required)	Neg.Wt.Ch.	-----	Elig.	-----	-----
(c)	0-540-A1C5 (Item 401(f) required)	Neg.Wt.Ch.	-----	Elig.	-----	-----
(d)	0-540-A1D5 (Item 401(f) required)	Neg.Wt.Ch.	-----	Elig.	-----	-----

Fig. 2-2   Sample specification sheet.

14

(e)	IO-540-C1B5 (Fuel injection installed per PAC Dwg. No. 23144) (Items 113, 114 (a) or 114 (b) and 401 (v) required).	Neg.Wt.Ch.	-----	Elig.	-----	-----
(f)	IO-540-D4A5 (Fuel injection installed per PAC Dwg. No. 24923) (Items 113, 114 and 401 (ak) or 401 (ba) required)	Use Actual Wt. Ch.	-----	-----	-----	Elig.
(g)	IO-540-N1A5	Use Actual Wt. Ch.	-----	-----	-----	Elig.
(h)	IO-540-R1A5 (Turbocharged - Items 122a required) Installed per PAC Dwg. 27137	Use Actual Wt. Ch.	-----	-----	-----	Elig.

Fig. 2-2   (continued)

Engine	Continental A-40, A-40-2, A-40-3, or A-40-4 (Also see item 305 and NOTE 4.)
Placard limits	A-40) Maximum, except takeoff – in. Hg., 2550 rpm (37 hp) A-40-2) A-40-3) Takeoff (one minute) – in. Hg., 2550 rpm (37 hp) A-40-4) Maximum, except takeoff – in.-Hg., 2575 rpm (40 hp) Takeoff (One minute) – in. Hg., 2575 rpm (40 hp)
Propeller	Maximum permissible diameter 81 in.
Fuel capacity	9 gals. (One tank in fuselage) (– 18)
Oil capacity	1 gal. (In engine crankcase)
No. passengers	1 (+ 9) and (+ 36)
Baggage	Landplane 20 lbs. (+ 44) Seaplane 12 lbs. (Placard compartment: "Includes anchor and rope 12 lbs. when carried."
Weights	Empty Use actual (Seaplane approximately 68 lbs. net increase over landplane.) Standard Landplane 1000 lbs. Seaplane 1070 lbs.
C.G. limits	Landplane (+ 10.6) and (+ 22.7) Seaplane (+ 10.2) and (+ 20.6) (Edo D – 1070 floats) (+ 11.8) and (+ 23) (Edo 54 – 1140 floats)
Specification basis	Approved Type Certificate No. 660
Serial Nos.	100 thru 1200, and 1999 and up mfrd. prior to 10 – 15 – 39 eligible. Approval expired as of that date. (See NOTE 4.)

Equipment: (Datum is wing leading edge) (*Means net increase)
Class I.

(a) Landplane

101. Propeller - wooden (fixed or adjustable pitch) 9 lbs.

102. 7.00 – 4 wheels (Shinn) and tires 21 lbs. (+ 3)

103. Tail skid   4 lbs.

(b) Seaplane   Item 101 PLUS

151. Edo D – 1070 float installation 145 lbs.

152. Aux. fin (on left and right stabilizer) (See NOTE 3)

153. Diagonal tension straps under front seat

Class II.

(a) Landplane

200. Miscellaneous items as noted in approved weight and balance report.

(b) Seaplane: Items 200 and 309

Class III.

(a) Landplane

301. Wheels

(a) 18 × 8 – 3 (Goodyear SLNBM Net decrease 2 lbs.

(b) 8.00 – 4 (Hayes 840) No weight change

(c) 8.00 – 4 (Hayes 841) with brakes 4 lbs. *

(d) Steerable tail wheel with 6 × 2.00 tire 3 lbs. *

(1) Aircraft Associates (Dwg. 1 – C)

(2) Aero Activities (DWG. 1 – C)

302. Battery (Reading 3 – BRL – 6)   8 lbs.

303. Carburetor heater   2 lbs.

304. Combination carburetor and cabin heater 3 lbs. (+ 39)

305. Dual ignition engine – Continental A – 40 – 5 3 lbs. * (– 39)
(Placard limits same as for A – 40 – 4)

306. Parachutes (a) Front   20 lbs. (+ 9)
(b) Rear   20 lbs. (+ 36)

307. Miscellaneous extra instruments

308. Emergency exit

309. Everel single blade propeller   12 lbs. *

310. Wheel streamlines   6 lbs.

311. 12 gal. fuel tank (replacing 9 gal. standard tank)

312. Frieze type balanced ailerons No wt. change

313. Skis   (Use actual weight)

(a) Piper S – 1000

(b) Air Transport A, 1220 – 480 or 1460 – 580

(c) Federal SA – 1, SA – 2, SC – 1 or SC – 2

(d) Marston MFS – 1200 or MFS – 1600

(e) Wash. Aircraft 1200

(f) Fairbanks MF – 5

(g) Aviation Service B

(h) Heath 665

314. Dual brake installation (Dwg. D4101 – C)

315. Two-hinge tail surfaces D – 4152 – C & D – 4157 – C.

(b) Seaplane: Items 302 to 308, inclusive, PLUS

351. Edo 54 – 1140 float installation (See NOTE 3)   147 lbs.

NOTE 1. Some aircraft of this model incorporate splices near the tip in the wing spars made at the factory.

NOTE 2. Eligible for export as follows, subject to inspection for equipment specified in Chapter XII of Inspection Handbook: (January 22, 1940)

(a) Canada – Landplane

Skiplane – not eligible. However, structure complies with Canadian requirements when item 313 (a) or equivalent is installed.

Seaplane – maximum standard weight 1070 lbs.

(b) Great Britain provided rear spars are reinforced in accordance with Piper Dwg. J2 – L15. If 13/16 in. spars are used they must also be reinforced in accordance with Piper Dwg. J3 – A – 107.

(c) Australia provided that:

(1) Wing spars are reinforced in accordance with Piper Dwg. J2 – L15. If 13/16 in. rear spars are used they must also be reinforced in accordance with Piper Dwg. J3 – A1107.

(2) The following placard speeds are displayed:
Level flight or climb 72 mph Ind.
Glide or dive 108 mph Ind.

Fig. 2-3   Sample aircraft equipment sheet.

(d)   All other countries.

NOTE 3.   Item 152 unnecessary when item 351 installed.

NOTE 4.   Serial No. 2309 also eligible as landplane with the following:

Engine   Franklin 4AC + 150 Series 40

Placard limits   Maximum, except takeoff — in. Hg., 1875 rpm (40 hp)

Takeoff (one minute) — in. Hg., 1875 rpm (40 hp)

Propeller   Maximum permissible diameter 79 in.

Placard speeds   Level flight or climb 78 mph Ind.

Glide or dive 108 mph Ind.

NOTE: Eligible for export to all countries except Canada, Great Britain and Australia.

(Formerly known as model J3F — 40. SPECIAL NOTE 11 on A.M.I.N. not pertinent.)

Fig. 2-3   (continued)

AIRCRAFT SERIAL NO. ___310L — 0100___ FAA REGISTRATION NO. ___N3250X___ DATE ___3 — 30 — 67___

1. Items of equipment marked (x) are installed when aircraft leaves factory.
2. Items marked (*) followed by a number replace an item of required equipment by that number.
3. Unless otherwise indicated, true values (not net change values) for weight and arm are shown.
4. Negative arms are distances forward of datum (forward face of the fuselage bulkhead just forward of the rudder pedals, Station 0.00).
5. Positive arms are distances aft of datum.
6. A separate FAA approval must be obtained if the following items are not installed per applicable Cessna drawings or accessory kit instructions.

A.	REQUIRED EQUIPMENT	DESCRIPTION	WEIGHT	ARM
1.	(x)	Altimeter (C661011 — 0201)	1.5	16.0
2.	(x)	Battery (Cessna 0511319 — 1)	46.0	45.0
3.	(x)	Belt, Pilot's Safety (Cessna S — 1618 — 3)	1.0	42.0
4.	(x)	Compass (CM2639 — 1)	1.0	15.0
5.	(x)	Engines (Continental IO — 470 — V Spec. 2)	846.9	0.0
6.	(x)	Filter, Engine Induction Air (Cessna 0850550 — 3)	1.5	11.0
7.	(x)	Fuel Pumps, Auxiliary Electric (Cessna 080420 — 1)	6.0	47.0
8.	(x)	Fuel Pumps, Engine Driven (Continental 630947 — 3)	4.0	11.0
9.	(x)	Gage, Manifold Pressure (Dual) (C662006 — 0101)	1.0	16.0
10.	(x)	Gage Unit, Left Engine Combination (CM2634 — L1)	1.5	16.0
11.	(x)	Gage Unit, Right Engine Combination (CM2634 — L1)	1.5	16.0
12.	( )	Alternators (50 Amp) (CMC 631111)	21.5	10.0
13.	(x)	Governors, Propeller (Woodward 201444)	6.5	— 17.0
14.	(x)	Indicator, Air Speed (CM3301 — 4)	0.5	16.0
15.	(x)	Indicator, Fuel Flow (Dual) (CM2650 — L4)	1.0	16.0
16.	(x)	Indicator, Stall Warning Audible (Safe Flight No. 285)	1.3	1.0
17.	(x)	Main Wheels, Brake Assemblies and Tires (6 Ply Rating, 6.50 x 10) Tube Type (Cessna 0841010 — 6)	66.0	55.0
18.	(x)	Nose Wheel and Tire (4 Ply Rating 6.00 x 6 Tube Type) (Cessna 0842210 — 2)	13.8	— 61.0
19.	(x)	Oil Radiator (Continental 626371)	18.0	— 17.0
20.	(x)	Pilot's Check List (0811720 — 1 & — 2)		
21.	(x)	Propellers (D2AF34C81/84JF — 3)	126.0	— 28.0
22.	(x)	Seat, Adjustable, Pilot (Cessna 0812202 — 101) (Inc. Seat Belt)	15.2	43.5
23.	(x)	Spinners, Propeller (Cessna 0855030 — 17) Blkhd (Cessna 0855030 — 14 & — 15)		
24.	( )	Tachometer (Dual) (CM2636 — L1)	1.5	16.0
25.	(x)	Voltage Regulators (50 Amp) (Continental 631656)	3.5	33.5

Fig. 2-4   Sample weight and balance equipment sheet.

The original weight and balance data and the equipment list are furnished by the manufacturer for a particular aircraft at the time of manufacture. Often this optional and required equipment is simply attached to the weight and balance data sheet. The equipment that is installed on the aircraft is marked with an (X) preceding the equipment item. Optional equipment that is not installed on the aircraft has an (0) or no mark in the ( ) preceding the item. (Fig. 2-4)

The excerpts of the equipment are quite similar to those found in the aircraft specifications with the weight and the arm given. This partial list is only the required equipment. Some items do not have an (X) preceding the item because it is not installed. If the list was complete it would be observed that optional equipment is used in lieu of the required equipment. It might also be noted that no net weights are used as was found in the aircraft specifications.

In addition to the weight and balance data and equipment list some manufacturers furnish *loading graphs* with this same information.

These loading graphs are to assist the pilot in determining the loading of the aircraft so that it may stay within the loaded center of gravity range.

There may be one graph for the total moment and total weight, or there may be separate graphs for moment and center of gravity. (Fig. 2-5)

The first chart represents a typical loading graph which would be used by a pilot. The items represented are pilot, passengers, fuel and baggage. The column on the left shows weight in pounds, and the numbers across the bottom show the moment in inch-pound times 1000. Looking at this graph it may be observed that 200 pounds of fuel will produce a moment of 10,000 if the moments of a load are added together with the basic empty moment and the load weights are added to the basic empty weight. We now have the total empty weight and total moment. This could be used to compute the loaded CG or graphs B and C could be used rather than a mathematical calculation.

Graph B shows the center of gravity loading envelope. The *envelope* is the heavy diagonal lines enclosed at the top. Notice that there is a smaller envelope for a *utility category* than for the *normal category*.

The total weight is given on the left side of the graph and the total moment is entered at the bottom. If two straight lines are drawn from these points and the lines intersect within the envelope, no further calculations are necessary by the pilot.

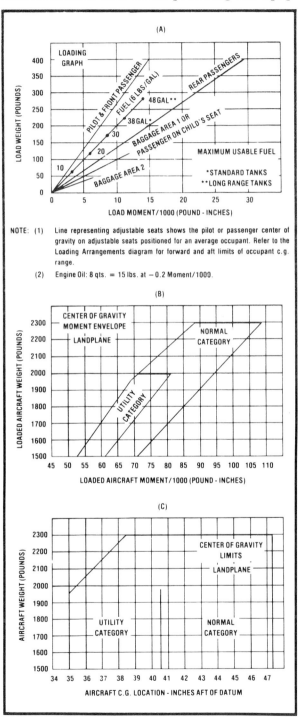

Fig. 2-5 *Typical loading graphs.*

19

Graph C is a pictorial graph of the CG limits of the aircraft. Different CG limits were given for different maximum weights and categories on the aircraft data sheets. By simply entering the weight on the left side and the CG from the bottom one can determine if the aircraft is within the loaded center of gravity range.

Another method that may be used are charts rather than graphs. These are nothing more than pre-calculated weights, arms, and moments that may make up a typical aircraft load.

One trend that we are seeing today is to include the equipment list and the loading information in the Operator's Manual (Flight Manual). This may include all of the information given in the weight and balance sheets or any portion of it. It will, however, always cover the aircraft loading of the aircraft. (Fig. 2-6)

The equipment list is very similar to other equipment lists except that this is *not* actually a list of the equipment installed on the particular aircraft. Instead, it is a list of what could be installed and what is required. The only items placed on the particular aircraft equipment list are what is installed (but it does not include the required equipment).

This list makes use of a code for various equipment. This is a suffix after the item number. The code is as follows:

R = Required equipment for certification.

S = Standard equipment items.

O = Optional equipment replacing required or standard items.

A. = Optional equipment in addition to required or standard items.

The maintenance manual will contain information on *jacking and leveling means* and will often include station locations. These station locations are used especially on large aircraft to locate equipment and are extremely helpful in determining the arm length when installing new optional equipment. (Fig. 2-7)

QUESTIONS:

1. In reference to Figure 2-1, the maximum weight in the utility category is:

   A.   2300 pounds.

   B.   2000 pounds.

   C.   2200 pounds.

   D.   It depends upon floats being installed.

2. In reference to Figure 2-1, what is the leveling means?

   A.   Upper door sill.

   B.   Firewall.

   C.   Top of the tail cone.

3. In reference to Figure 2-1, what is the unusable fuel for the 172M?

   A.   21 gallons at (+48).

   B.   18 pounds at (+46).

   C.   24 pounds at (+46).

4. In reference to Figure 2-1, on Aircraft #17267587, what is done with the oil?

   A.   Drain the oil.

   B.   Subtract the undrainable oil.

   C.   Fill the oil tank before weighing.

5. In reference to Figure 2-2, which PA24 series does not require a stall warning indicator?

   A.   PA24.

   B.   PA24-250.

   C.   PA24-260.

ITEM NO	EQUIPMENT LIST DESCRIPTION	REF DRAWING	WT LBS	ARM INS
	B. Landing Gear & Accessories			
B01 – R	Wheel, Brake & Tire Assy, 6.00X6 Main (2)	C163015 – 0201 OR	41.7*	57.8*
		C163015 – 0206		
	Wheel Assy., McCauley D30260 (Each)	C163003 – 0101 OR	6.4	58.2
		C163015 – 0102		
	Brake Assy., McCauley (Left)	C163032 – 0111	1.9	54.5
	Brake Assy., McCauley (Right)	C163032 – 0112	1.9	54.5
	Tire, 4 – Ply Blackwall (Each)	C262003 – 0101	8.5	58.2
	Tube (Each)	C262023 – 0102	1.8	58.2
B04 – R	Wheel & Tire Assy., 5.00 X 5 Nose	C163015 – 0109	9.3*	– 6.8*
	Wheel Assy., McCauley	C163003 – 0401	3.0	– 6.8
	Tire, 4 – Ply Blackwall	C262003 – 0102	4.7	– 6.8
	Tube	C262023 – 0101	1.2	– 6.8
B10 – S	Fairing Installation, Wheel (Set of 3)	0541225 – 1	17.8*	47.1*
	Nose Wheel Fairing		4.0	– 4.9
	Main Wheel Fairing (Each)		5.7	60.3
	C. Electrical System			
C01 – R	Battery, 12 Volt, 25 Amp Hour	0511319	23.0	0.0
C04 – R	Regulator, 14 Volt, 60 Amp Alternator	C611001 – 0201	0.5	3.5
C07 – A	Ground Service Plug Receptacle	0501053	2.7	– 2.6
C16 – 0	Heating System. Pitot (Net Change)	0422355	0.6	24.4
C22 – A	Lights, Instrument Post	0513094	0.5	16.5
C25 – A	Light, Map (Control Wheel Mounted)	0570087	0.2	21.5
C28 – S	Light, Map & Instrument Panel Flood (Doorpost Mounted)	0700149	0.3	32.0
C31 – A	Lights, Courtesy Entrance (Set of 2)	0521101	0.5	61.0
C40 – A	Detectors, Navigation Light (Set of 2)	0701013 – 1, – 2	NEGL	– –
C43 – A	Light Installation, Omniflash Beacon	0506003	2.1*	184.2*
	Beacon Light On Fin Tip	C621001 – 0103	0.4	243.0
	Flasher Power Supply	C594502 – 0101	0.8	205.8
	Resistor (Memcor)	OR95 – 1.5	0.3	208.1
C46 – A	Light Installation, Wing Tip Strobe	0501027	3.4*	43.3*
	Flasher Power Supply (Set of 2 in Wing)	C622007 – 0101	2.3	47.0
	A. Powerplant & Accessories			
A01 – R	Engine, Lycoming 0 – 320 – E2D (Includes All Electric Starter and			
	Vacuum Pad)	0550319	266.0*	– 20.0*
A05 – R	Filter, Carburetor Air	C294510 – 0301	0.5	– 26.0
A09 – R	Alternator 14 Volt, 60 Amp (Belt Drive)	C611501 – 0102	10.8	– 29.0
A17 – R	Oil Cooler Installation	0550319	2.5*	– 2.5*
	Oil Cooler (Harrison Or	8526250	2.1	– 2.5
	Stewart Warner)	8406E	2.1	– 2.5
A21 – A	Filter Installation, Full Flow Engine Oil	1756004 – 1	4.5*	– 5.0*
	Adaptor Assy (Lycoming)	755288	0.8	– 6.5
	Filter Can Assy (AC)	6436992	1.8	– 3.3
	Filter Element Kit (AC)	6435683	0.3	– 3.3
A33 – R	Propeller Assy. (Fixed Pitch – Landplane)	C161001 – 0306	35.9*	– 38.5*
	Propeller (McCauley)	1C160/DIM7553	30.1	– 39.1
	3.5 Inch Prop Spacer Adaptor (McCauley)	C4516	3.6	– 35.4
A33 – 0	Propeller Assy. (Fixed Pitch – Floatplane)	C161001 – 0307	37.2*	– 38.6*

Fig. 2-6   Sample page from Operator's Manual of a Cessna C-172M.

Fig. 2 — 6 Continued

	Propeller (McCauley)	1A175/ETM8042	31.5	− 39.1
	3.5 Inch Prop Spacer Adaptor (McCauley)	C4516	3.6	− 35.4
A41 − R	Spinner Installation, Propeller	0550320	2.0*	− 41.4*
	Spinner Dome	0550236 − 8	1.2	− 43.1
	FWD Spinner Bulkhead	0550321 − 4	0.3	− 40.8
	Aft Spinner Bulkhead	0550321 − 1	0.4	− 37.3
A61 − S	Vacuum System Installation	0501055 − 2	4.3*	− 3.0*
	Dry Vacuum Pump (Av Wt of 4 Types)	C431003 −	2.8	− 6.3
	Filter	C294502 − 0101	0.2	4.7
	Vacuum Gauge	C668509 − 0101	0.1	16.2
	Relief Valve − Regulator	C482001 − 0401	0.5	4.5
A70 − A	Primer System, Engine Three Cylinder	1701015	0.5	− 12.0
A73 − A	Oil Quick Drain Valve (Net Change)	1701015	0.0	--

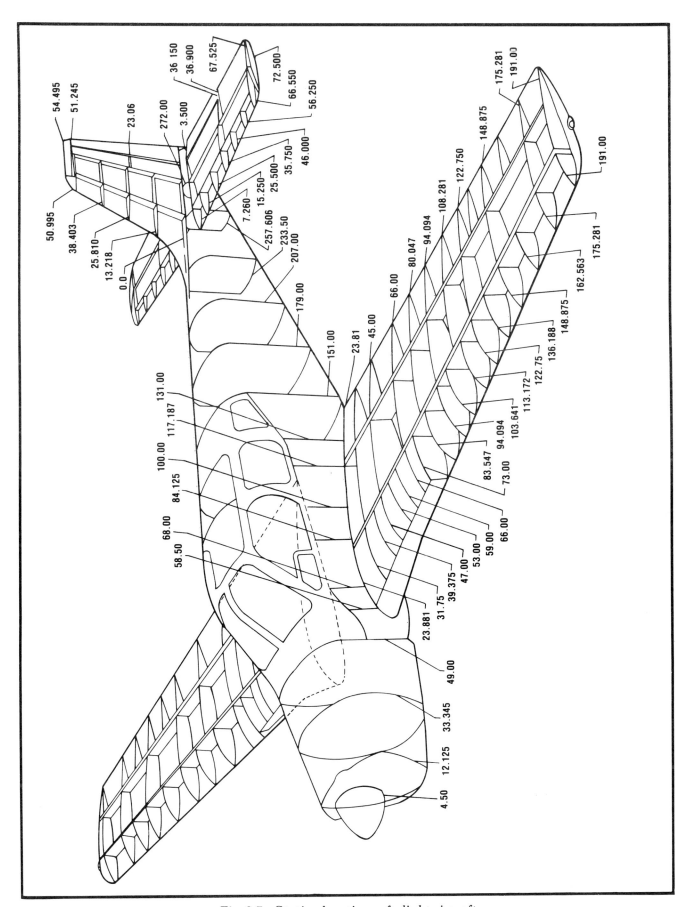

Fig. 2-7   Station locations of a light aircraft.

23

6. In reference to Figure 2-2, what does the exhaust gas temperature gauge weigh in Item 607 (c)?

   A.   2 pounds.

   B.   2 pounds is the net weight, not the actual weight.

7. In reference to Figure 2-2, what does Item 607 (a) weigh?

   A.   1 pound.

   B.   1 pound is the net weight, not the actual weight.

8. In reference to Figure 2-2, Item 609 (f) has no arm measurement. What should be done?

   A.   Reweigh the aircraft.

   B.   It is located at Station (0).

   C.   Add the actual weight. The arm is the same.

9. In reference to Figure 2-3, the Class I equipment is:

   A.   Required.

   B.   Optional.

   C.   May be either required or optional.

10. In reference to Figure 2-3, if Item 309 was used to replace Item 101, what would be the weight change?

    A.   9 pounds.

    B.   12 pounds.

    C.   Use the actual weight.

11. In reference to Figure 2-5A, what is the moment for 100 pounds of baggage?

    A.   10.

    B.   1000.

    C.   10,000.

12. In reference to Figure 2-5B, if the aircraft was loaded in such a manner that the weight was 1900 pounds and the moments were 8500 which category may the aircraft be operated under safely?

    A.   Normal.

    B.   Utility.

    C.   Both.

13. In reference to Figure 2-5C, if the empty weight was 1900 pounds and the CG was 40 inches, what category may the aircraft be operated under?

    A.   Utility.

    B.   Normal.

    C.   Either normal or utility.

14. In reference to Figure 2-4, why is Item 12 not installed when it is required equipment?

    A.   It is not required.

    B.   Optional alternators are used.

    C.   The engine will run without it.

15. In reference to Figure 2-6, is Item C01-R installed on the aircraft?

    A.   Yes.

    B.   No.

    C.   Yes, if an alternate is not used.

16. In reference to Figure 2-6, what type of item is A73A?

    A. Optional.

    B. Standard.

    C. Additional.

17. In reference to Figure 2-6, may items with suffix A replace items with the suffix S?

    A. Yes.

    B. No.

18. Where would jacking procedures be found for a particular aircraft?

    A. Operator's Manual.

    B. Maintenance Manual.

    C. Aircraft Specification Sheet.

19. Where would the leveling means be located for a particular aircraft?

    A. Data Sheet.

    B. Maintenance Manual.

    C. Both A & B.

20. Station locations are numbered from the nose to the tail.

    A. True.

    B. False.

CHAPTER III

Weighing the Aircraft

Fig. 3-1   Leveling aircraft on scales prior to weighing.

Weighing the aircraft is the most crucial step in a weight and balance calculation because all of the loading of the aircraft that is done is based on these figures. It might also be noted that most additions and subtractions of equipment are done mathematically based on the figures derived from the basic calculations.

It is the tendency for aircraft to become heavier as they become older. It is for this reason that aircraft used for charter are reweighed periodically while other aircraft in private use may not be reweighed in the lifetime of the aircraft.

A. *Equipment*

1. *Scales*

All aircraft should be weighed in a closed hangar with a fairly level floor. If the aircraft were to be weighed outside, the wind over the wings would adversely affect scale readings, thus giving *lighter* readings than the actual aircraft weight.

Difficulty will be encountered in leveling the aircraft if the floor is not fairly level, and a possibility exists of placing a side load on the scales which also results in erroneous scale readings.

Usually, at least three scales are used during the actual weighing of the aircraft. On some of the larger aircraft four scales are required with two of these on the nose of the aircraft and two scales on the main gear. Many light aircraft have been weighed with two scales with a bridge across the main gear and a few have been weighed with one

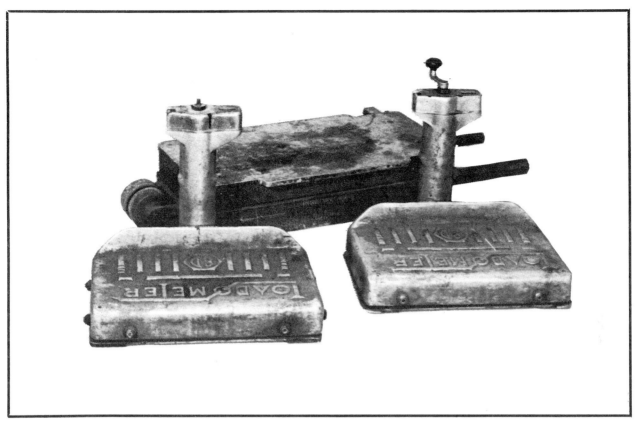

Fig. 3-2   Low platform scales.

scale. Neither of the latter methods are recommended because of the possibility of damage to the aircraft and inaccurate readings due to side loading of the scale.

The type of scales which are used will vary from shop to shop and the size and type of aircraft to be weighed. In general terms there are *two* types of scales which are commonly used. They are the *mechanical platform* type and the *electronic* type.

Usually shops dealing with general aviation use the mechanical platform scales while larger aircraft are weighed with the electronic scales.

In both cases, the scales should be of high quality and should be checked for accuracy at least every 24 months (Figs. 3-2 and 3-3).

The mechanical platform scales should be of a low profile type to ease getting the aircraft onto the platform. They should be capable of loads in excess of the expected weight (usually by 150%). This is recommended so that excessive loads plac-

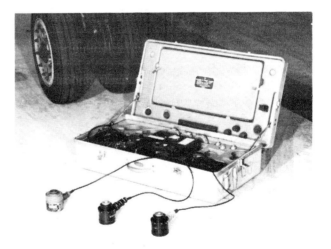

Fig. 3-3   Electronic scales.

ed on the scales during movement of the aircraft on or off the platform will not cause damage to the mechanism. If the scale range is too great, the accuracy in the lower ranges might be affected.

Fig. 3-4   Load cells being placed on jacks.

Most heavy aircraft today are weighed by electronic scales. These are often referred to as *load cells* or *strain gauges*. These load cells reflect a voltage change for the weight imposed upon them. This variation in voltage is changed into a weight reading. In most instances the load cells are placed between the jack pads and the jacks in order to obtain the weight (refer to Fig. 3-4). Load cells are also available for axle jacks and ramps for our larger jet aircraft. These utilize more than one main wheel (Fig. 3-5). When load cells are used, special precautions should be observed. Do not load the rim of the cell, and be sure that the jacking is done evenly because the rim loading or the side load that can be imposed, can destroy the load cell accuracy.

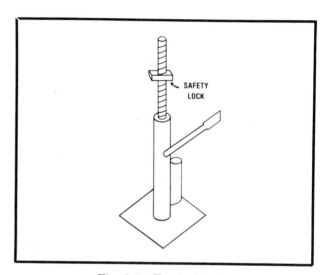

Fig. 3-6   Typical jack.

2.   Jacks

Jacks are often necessary to weigh various aircraft. If electronic weighing equipment is used, the aircraft must be jacked into position in order to place the weight on the load cell. On tail-wheel aircraft, the tail must be raised to a level attitude. Many manufacturers recommend that the aircraft be weighed from the *jack points*. When jacks are used it is recommended that they have a capacity of 150% of the load to be carried, fit well in the jack pad, and also be equipped with safety locks. All jacking should be performed evenly and strictly in accordance with the manufacturers' recommendations. This is because some of our aircraft today require ballast in the nose or tail during jacking operations (Fig. 3-6 and 3-7).

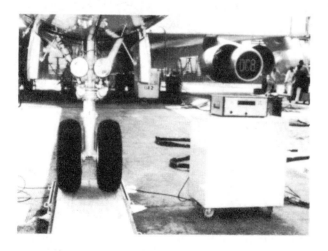

Fig. 3-5   Ramp type electronic weighing system.

Fig. 3-7   Typical proper jacking operation.

a.   Tare

Tare is the additional-weight items that are used during the weighing procedure. This could be the chocks used to hold the wheels on the scale platforms because brakes are never applied during weighing due to possible side loading of the scale. Tare might also be a jack placed on the scale platform or ballast required for the jacking operation. Regardless of what the tare may be, it must be subtracted from the scale reading before empty weight and the center of gravity are computed (Fig. 3-8).

3.   Additional weighing equipment

Other equipment that will be necessary for weighing the aircraft are: a spirit level, plumb bobs, measuring tape and chalk or soapstone.

The spirit level may be used for leveling the aircraft longitudinally if provisions are made for the use of a spirit level. Normally, light aircraft have these type of provisions while larger aircraft use the plumb bob and leveling scale method. Most

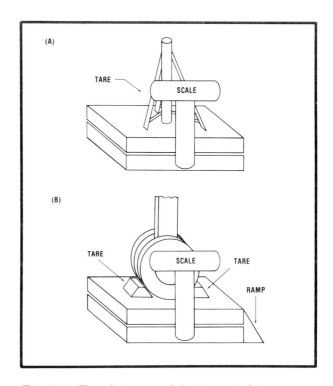

Fig. 3-8   Tare being used during weighing operation.

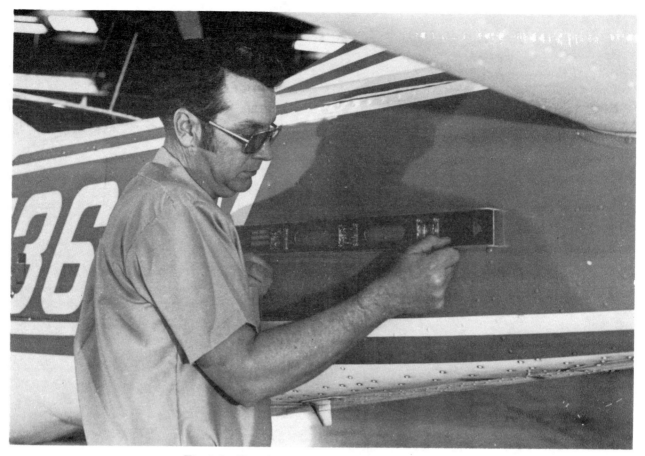

Fig. 3-9   Leveling aircraft prior to weighing.

light aircraft are leveled only in the longitudinal direction for weighing. If it is possible to check the lateral level, it would be advisable. But unless the aircraft is out of level a great amount laterally, it will have very little effect on the scale readings. *However, the longitudinal leveling is most critical to the scale reading.*

The plumb bob may be used with the leveling scale. It will also be used for dropping points to the floor for such items as datum lines and weighing points so actual measurements may be taken for computation.

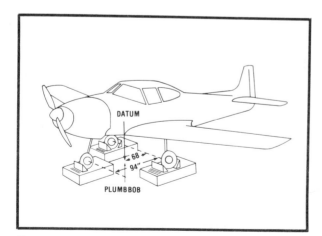

Fig. 3-10   Use of plumb bob for datum measurement.

Fig. 3-11   Plumb bob measurement from the nose wheel center line.

30

Fig 3-12 Positioning aircraft seats prior to weighing.

B. Preparation of the Aircraft

The airframe should be thoroughly cleaned before weighing. A little dirt spread over the large area of the aircraft will make quite a difference in the weight. This cleaning should include not only the outside, but the interior, wheel wells, and the baggage compartments. All items not considered as equipment should be removed. Aircraft always have a tendancy to become heavier as the aircraft becomes older. This is due to dirt in inaccessible places, paint, and items added of negligible weight.

Fuel should be drained from the aircraft tanks before weighing. The fuel remaining is considered residual fuel and is considered part of the empty weight. Generally, on light aircraft, the fuel is drained from the fuel sumps or screens of the system.

3-13 Draining fuel from the tank sumps.

Always check the aircraft specifications or data sheets. Unusable fuel is considered a part of the empty weight just as undrainable fuel is considered a part of the empty weight, therefore a certain amount may have to be added to the tank after draining it, or it will have to be computed after the CG is established from the scale readings. *This fuel is a part of the empty CG and weight.*

Under special circumstances it may not be possible to drain the fuel in the aircraft. In these cases, the tanks should be filled, with the weight of the fuel subtracted from the weight of the aircraft. This method is not advisable on light aircraft, and should be used on large aircraft *only* when being tested with a hydrometer to determine the actual weight of the fuel. This is especially important with jet fuel where the fuel density varies so greatly.

Oil should be removed from the engine or oil tank. Like the fuel, oil trapped in the system is called the residual oil and is considered a part of the empty weight. Unusable oil is not a part of the empty weight. When it is not practical to drain the oil, the tanks are filled and subtracted from the empty weight. The oil density will not vary to the extent that a hydrometer reading would be required.

Fluids in the other systems of the aircraft will normally be full. This includes hydraulic fluid, constant speed drive oil, and anti-icing fluid. Fluid such as galley water and toilets should be drained. In all cases the final determination should be made in accordance with the manufacturer's recommendations.

The position of the control surfaces, such as flaps, are usually in the *up* position. This position will not affect the weight of the aircraft, but will have an effect on the center of gravity. On certain helicopters, the position of the main rotor can affect the center of gravity location which is quite critical when the center of gravity range may be no more than three inches. Again, the manufacturer's recommendations should be followed in regards to the position of all control surfaces.

C. Positioning the Aircraft

If an electronic weighing system is to be used, no further positioning of the aircraft will be necessary because the load cells will be placed be-

Fig. 3-14   Measuring the distance from the nose wheel to the main wheel.

tween the jack and the jackpoint. However, on most light to intermediate aircraft *low profile* mechanical scales are used and the aircraft is usually rolled onto these scales using ramps. If this method is used, it is advisable to use chocks on the scales both fore and aft so that the aircraft does not roll during the weighing procedure. Parking brakes should never be used because the loads that may be placed on the scales will destroy the accuracy of the readings.

When a tail wheel aircraft is to be weighed, additional positioning will be required because the tail will have to be raised to the flight attitude to level the aircraft. In many cases this is at least six feet making it quite difficult for a man to reach. The possibility also exists of raising the tail too high so that the aircraft falls on its nose. To avoid such catastrophies, it is suggested that a strap be placed over the tail section with a man holding the strap on each side during the raising and lowering operation.

Place the scale as close to the tail wheel as possible. Such devices as placing the tail on a freight elevator or placing a platform on a fork lift have been used with success if the location of the tail wheel and the tail itself allows this method to be accomplished. In many cases a tail stand or jack must be used with the scale placed under the stand or the jack.

Leveling information can be obtained in the aircraft data sheets or in the manufacturer's manual. The leveling means and its location vary from aircraft to aircraft. If the aircraft is weighed on jackpoints, the jacks are simply raised or lowered in order to reach the longitudinal level position. When the aircraft is weighed with scales under the wheels, the level position can be obtained on nose wheel aircraft by raising or lowering the nose wheel strut. The tail wheel aircraft, however, present a problem if some moveable means is not used to raise and lower the tail. This may necessitate lifting the aircraft off the stand several times which should be avoided if at all possible. On at least one of the older large tail wheel aircraft provisions were made by means of a chart in order to weigh the aircraft with the tail down.

Remember that when items such as chocks, jacks and stands are used in the weighing operation on top of the scales, their weight is considered tare. Tare must always be subtracted from the scale readings. It is suggested that the tare weight be weighed on scales of a lower range because the accuracy of the aicraft scales is likely to be inaccurate in the lower range readings.

D.   Measurements

On most aircraft actual measurements will have to be taken to locate the datum and the mea-

Fig. 3-15   Recording the scale readings.

surements from the datum to the weighing points. For aircraft where the datum is located ahead of the aircraft, a reference point on the aircraft is usually given in the maintenance manual from which a plumb bob can be dropped to locate the datum. When the datum passes through the aircraft, the plumb bob is simply dropped from that point. Once the datum is located on the floor, the plumb bob is simply suspended from the weighing points and the measurements are taken. It is advisable to draw a line between the main gear or jack points so that all measurements are taken from the center line of the aircraft (Fig. 3-16).

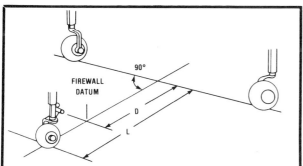

Fig. 3-16   Typical measurements taken during weighing operation.

In a few instances this information may be found in the operator's manual or in the maintenance manual where the station locations are given.

1.   Basic computation

Since the computations of the weight and balance are required on all aircraft and all changes must be noted in the aircraft logbooks, most shops will record the scale readings on a form which will become a permanent part of the aircraft's records. By using such a form, common errors are less likely to occur during computation. It is also advisable to sketch the aircraft weighing points and distances when working the computation. This will enable you to see your work as the problem is solved.

The formulas used in computing the center of gravity are varied. No standards have been established at this time. Although most manufacturers use the same basic formulas, they use different letter designations for the item. Some manufacturers may refer to A as the distance from the main gear to the nose wheel while another may refer to this point as C.

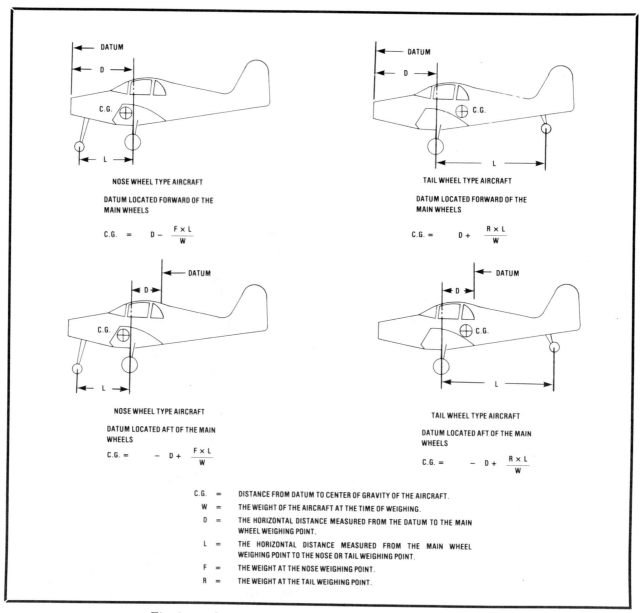

NOSE WHEEL TYPE AIRCRAFT

DATUM LOCATED FORWARD OF THE
MAIN WHEELS

$$C.G. = D - \frac{F \times L}{W}$$

TAIL WHEEL TYPE AIRCRAFT

DATUM LOCATED FORWARD OF THE
MAIN WHEELS

$$C.G. = D + \frac{R \times L}{W}$$

NOSE WHEEL TYPE AIRCRAFT

DATUM LOCATED AFT OF THE MAIN
WHEELS

$$C.G. = -D + \frac{F \times L}{W}$$

TAIL WHEEL TYPE AIRCRAFT

DATUM LOCATED AFT OF THE MAIN
WHEELS

$$C.G. = -D + \frac{R \times L}{W}$$

C.G. = DISTANCE FROM DATUM TO CENTER OF GRAVITY OF THE AIRCRAFT.

W = THE WEIGHT OF THE AIRCRAFT AT THE TIME OF WEIGHING.

D = THE HORIZONTAL DISTANCE MEASURED FROM THE DATUM TO THE MAIN WHEEL WEIGHING POINT.

L = THE HORIZONTAL DISTANCE MEASURED FROM THE MAIN WHEEL WEIGHING POINT TO THE NOSE OR TAIL WEIGHING POINT.

F = THE WEIGHT AT THE NOSE WEIGHING POINT.

R = THE WEIGHT AT THE TAIL WEIGHING POINT.

Fig. 3-17   Center of gravity formulas for various aircraft.

One formula used quite extensively today is contained in the FAA Advisory Circular 43.13.1A. This system utilizes four separate formulas. The user selects one of these formulas depending upon the weighing points and the datum location in reference to the weighing points. This system has some advantages and some disadvantages. The advantage is that there are less chances of error in negative and positive numbers. The disadvantage is that there are four formulas with different configurations to remember. (Fig. 3-17)

The other formula which is used extensively is the *total moment divided by total weight* for-mula. This is used exclusively when equipment is added or subtracted from the aircraft. The advantage is that there is only one formula to remember. However, there is the possibility of error where negative and positive moments exist if the datum is not located ahead of, or at the nose of the aircraft.

For the purposes of this publication, the formulas from AC 43.13.1A will be used for all calculations for the empty center of gravity. For the addition or subtraction of equipment the total moment divided by total weight formula will be used.

2. Mean aerodynamic chord

The *mean aerodynamic chord* (MAC) is the measurement used to determine the location of the center of gravity. However most manufacturers of light aircraft indicate the center of gravity in inches. The percentage of mean aerodynamic chord is used almost exclusively on large aircraft. This measurement is an imaginary straight line from the leading edge of the wing of the average airfoil section to the trailing edge. (Fig. 3-18)

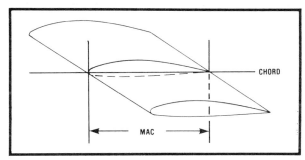

Fig. 3-18   MAC measurement.

This measurement is usually expressed as a percentage (although it may be expressed in inches).

To determine the MAC on a typical sweptback wing, it would be measured chordwise at the *root and the tip*. These measurements would be used to determine the average airfoil section. (Fig. 3-19)

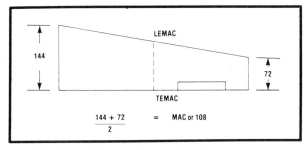

Fig. 3-19   Tapered wing MAC measurement.

For example: the root measurement of a certain wing is 144 inches and the tip is 72 inches. The MAC would be 108 inches. The leading edge of the mean aerodynamic chord is abbreviated LEMAC. The trailing edge of the mean aerodynamic chord is abbreviated TEMAC. The center of gravity will always lie between LEMAC and TEMAC if the aircraft is within CG limits. The center of gravity is expressed as a percentage and is actually —inches behind LEMAC (Fig. 3-19).

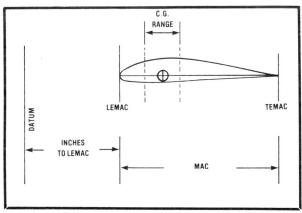

Fig. 3-20 Center of gravity falls between LEMAC and TEMAC.

Since all calculations are taken from the datum, we may determine that the center of gravity lies 280 inches aft of the datum, and LEMAC is 240 inches, and TEMAC is 320 inches, and the length of the MAC is 320″ - 240″ or 80 inches. (Fig. 3-21)

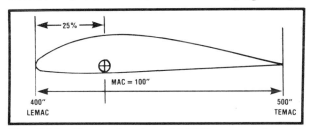

Fig. 3-21   Determination of MAC center of gravity.

If the center of gravity is expressed as a precentage, then the center of gravity location can be determined by TEMAC − LEMAC, then multiplied by the CG in percentage.

For example: LEMAC = 400″ TEMAC = 500″ CG is 25% MAC.

$$TEMAC = 500$$
$$LEMAC = \underline{400}$$
$$MAC \quad = 100$$
$$MAC = 100 \times 25\% = 25 \text{ inches}$$

To determine the percent of MAC the following formula will be used:

$$\% \text{ of MAC} = \frac{\text{Distance from LEMAC} \times 100}{\text{MAC}}$$

For example: MAC = 170″ LEMAC = 187″ CG = 207.4″

$$CG \, 12\% = \frac{(207.4 - 187) \times 100}{170}$$

35

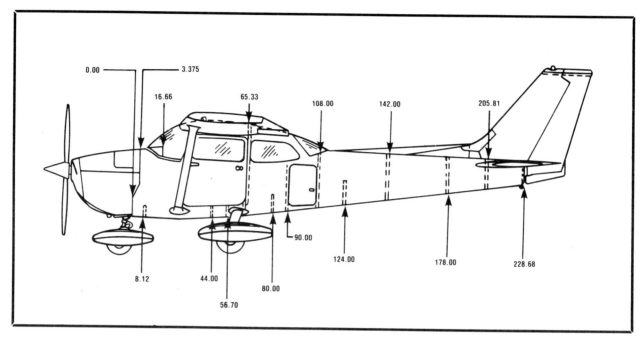

Fig. 3-22 View of Cessna C-172M.

E. Nose Wheel Aircraft

The nose wheel aircraft is by far the most popular aircraft flown today in general aviation and in the transport category. This is mainly due to its ease in handling and because of the improved runways that are available today. For this reason, the most often weighed aircraft will be the nose wheel type. This is because of the FAR 135 regulation that requires physically re-weighing the aircraft every three years.

Below are two weight and balance calculations for nose wheel aircraft.

PROBLEM 1 N

Our first problem will be a Cessna 172 M.

The following information is obtained from the aircraft data sheet.

Datum: Front face of firewall.

Leveling Means: Upper door sill.

CG Range:

Normal Category: (+40.8) to (+46.4) at 2200 pounds
(+36.4) to (+46.4) at 1733 pounds

Utility Category: (+38.4) to (+40.3) at 1950 pounds
(+36.4) to (+40.3) at 1833 pounds

Empty CG Range: None.

No. of Seats: 4 [(2 at (+36)] and 2 at (+70).

Maximum Baggage: 120 pounds at (+95).

Fuel Capacity:

39 gallons total, 36 gallons usuable.

[two 19.5 gallon tanks in wings at (+48)]

See Note 1 for weight of unusable fuel and oil.

Maximum Weight:

Normal: 2300   Utility: 2000

Oil Capacity: 2 gallon (−14) includes 1 gallon unusable.

*Note 1: Empty weight must include 24 pounds at (+46) and full oil of 11.3 pounds at (−14).

*There seems to be a trend that has recently started to include the oil in the certificated empty weight. Cessna refers to this as basic empty weight.

View of 172.M: See Fig. 3-22.

36

Step 1: Clean the aircraft and remove all loose articles and drain only the fuel. Check oil level. (Remember the 24 pounds of fuel must be included.)

Step 2: Place the aircraft in a closed hangar.

Step 3: Check the scale adjustments for zero and place the aircraft on the scales. Use chocks under the wheels. Use no brakes.

Step 4: Place a spirit level on the upper door sill and level the aircraft using the nose strut.

Step 5: Record the scale readings and subtract the tare (chocks).

	Gross	Tare	Net
Left Main	476	2	474
Right Main	471	2	469
Nose Wheel	402	0	402
	Total Empty Weight:		1345 pounds

Step 6: Using a plumb bob and steel tape, measure the horizontal distance from the nose wheel to the main wheels and the datum to the nose wheel. The distances are (68.75) and (−8.5) respectively.

Step 7: Using the formula for nose wheel aircraft, compute the empty CG:

$$D - \frac{F \times L}{W} \qquad 60.25 - \frac{402 \times 68.75}{1345} = 39.7$$

Note: 68.75 minus 8.5 will give D for the formula (60.25). Our empty CG is 39.7 aft of the datum.

Step 8: Using the formula: CG × weight = moment, compute the moment: 39.7 × 1345 = 53396.5.

Step 9: In order to obtain the basic empty weight as mentioned in Note 1, fuel must be added. The oil was left in the engine. Therefore no added computation for oil is necessary. The unusable fuel was 24 pounds at (+46). Using the total moment over total weight formula, the fuel may be added. The moment for the fuel is weight × arm or 24 × 46 = 1104. The total weight is now 1345 + 24 or 1369.

$$\text{The CG} = \frac{54500.5}{1369} = 39.8 \text{ inches.}$$

The basic empty weight is 1369.

The basic empty CG is 39.8″. The basic empty moment is 54500.5.

PROBLEM 2 N

For the second nose wheel problem a Piper PA 24-250 (better known as a Piper Comanche), serial no. 24-26M, will be used. In order to obtain information in the aircraft specifications the number designation must be used. From the aircraft specifications the following information is obtained:

Datum: 79″ forward of the wing leading edge at wing station 65.5 (intersection point of tapered edge with straight leading edge).

Leveling Means: Level from two rivnuts located right side above baggage door.

Center of Gravity: Serial No. 24-2299 up.

CG Range: (+86.0) to (+93.0) at 2900 pounds

(+82.5) to (+93.0) at 2600 pounds

(+80.5) to (+93.0) at 2000 pounds or less.

Empty Weight: None.

Empty CG: None.

Maximum Weight: 2900 pounds Serial No. 24-2299 and up.

Number of Seats: 4 [2 at (+85) and 2 at (+118.5)].

Maximum Baggage: 200 pounds (rear compartment) (+142).

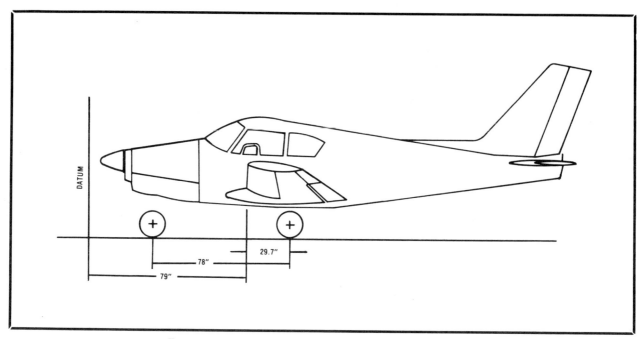

Fig. 3-23   View of Piper PA24-250 (Piper Comanche).

Fuel Capacity: 56 gallons (two 28 gallon wing tanks) (+90).

Oil Capacity: 3 gallons (+28).

View of PA24-250: See Fig. 3-23.

Step 1: Clean the aircraft and remove all loose articles from the interior and drain the fuel and oil.

Step 2: Place the aircraft in a closed hangar.

Step 3: Check the scale adjustments for zero and place the scales under the landing gear. *Note:* (parking brakes *OFF*)

Step 4: Place spirit level on the screws inserted in the leveling points and adjust for level using the nose strut.

Step 5: Record the scale readings and subtract tare (chocks).

	Gross	Tare	Net
Left Main	570	5	565
Right Main	566	5	561
Nose Wheel	588	3	585
Total Weight:			**1711 pounds**

Step 6: Using a plumb bob and steel tape, measure the horizontal distance from the nose wheel center line to the main wheel center line and the datum to the main wheel center line. (78 inches and (108.7 inches) respectively.

Step 7: Using the formula for nose wheel aircraft compute the empty CG.

$$CG = D - \frac{F \times L}{W} \qquad 108.7 - \frac{585 \times 78}{1711} = 82.0$$

Step 8: To determine the total empty moment, use the following formula:

$$CG \times Weight = Moment$$

$$82 \times 1711 = 140302$$

Step 9: The useful load of the aircraft is the empty weight subtracted from the maximum weight.

$$2900 - 1711 = 1189 \text{ pounds}$$

$$Useful \ Load = 1189 \text{ pounds}$$

F. Tail Wheel Aircraft

The new tail wheel aircraft built today are usually special purpose aircraft. Usually they are used to

38

fly in and out of unimproved landing areas where it would not be practical to operate a nose wheel type aircraft. Because of the work that is required of these aircraft, they are built to carry heavy loads for such jobs as agriculture and hauling passengers, supplies and equipment into isolated areas. Many of this type of aircraft will operate in either of a restricted category or under FAR Part 135 which are air taxi regulations. The aircraft used for taxi must be physically re-weighed every 36 months.

Below is a weight and balance calculation for a tail wheel aircraft.

PROBLEM 1 T

This problem will be a Cessna 188B (better known as Ag Truck), serial no. 18802965T, which may operate in either the normal or restricted category.

From the aircraft specification data sheet the following information is obtained.

Datum: Fuselage station (0.0). Front face of firewall.

Leveling Means: Two jig located nutplates and screws installed on the left tail cone.

CG Range: (+39.0) to (+45.5) at 2300 pounds or less. Normal Category: (+41) to (+45.5) at 3300 pounds.

Empty CG Range: None.

Maximum Weight: 3300 pounds (Normal Category)

No. of Seats: 1 at (+91) to (+95).

Maximum Cargo: 1670 pounds at (+43.0 Station. With letter *T* suffix on Serial No. (1800 pounds) at (+43) Station.

Additional Limitations for Restricted Category:

CG Range: (+39) to (+47.5) at 2300 pounds or less

(+39.4) to (+47.5) at 2500 pounds

(+41.0) to (+46.4) at 3300 pounds

(+39.3) to (+45.2) at 4200 pounds

Maximum Weight: Serial Numbers: 1880135T and up — 4200 pounds

View of Cessna 188B: Refer to Fig. 3-24.

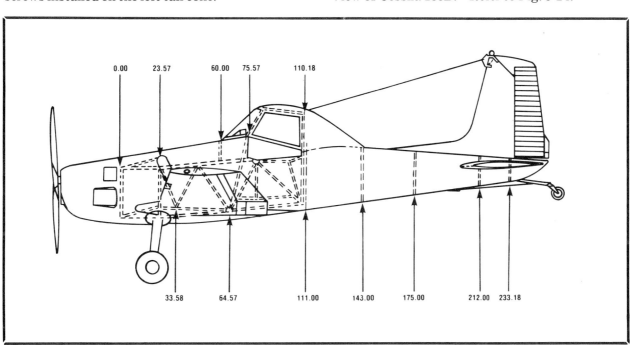

Fig. 3-24 View of Cessna 188B Ag Truck.

Step 1: Clean aircraft and remove loose articles (check hopper) and drain the fuel and oil.

Step 2: Move the seat to the most forward position, control surfaces in neutral and retract the flaps.

Step 3: Place the aircraft in a closed hangar.

Step 4: Check the scales for zero and place the scales under the main gear and tail wheel using a jack between the scale and tail wheel. (Follow precautions mentioned previously.)

Step 5: Place a spirit level on the leveling lugs and level the aircraft using the tail jack.

Step 6: Record the scale readings and subtract tare (chocks and tail jack).

	Gross	Tare	Net
Left Wheel	1123	5	1118
Right Wheel	1120	5	1115
Tail Wheel	237	20	217
Empty Weight:			2450 pounds

Step 7: Using a plumb bob and a steel tape, measure the horizontal distance from the datum to the main gear center line and the main gear center line to the tail wheel center line. They should be (17.5 inches) and (245.5 inches) respectively.

Step 8: Using the following formula for tail wheel aircraft compute the empty CG:

$$CG = D + \frac{R \times L}{W} \qquad 17.5 + \frac{217 \times 245.5}{2450} = 39.25$$

Empty CG = 39.25 inches aft of the datum.

Step 9: To locate the total moment: multiply the CG × weight. 39.25 × 2450 = 96162.5 = Empty Moment

G. Helicopter

Most helicopters will be weighed very similarly to fixed wing aircraft. If the helicopter is equipped with skid gear, the jack points will be used as the weighing points. Normally this will provide three points for weighing just like a tail wheel aircraft. The position of the rotor may be quite critical at the time of weighing and can adversely affect the scale reading if positioned incorrectly. Always refer to the manufacturer's recommendations for this information.

Another unique feature of helicopter weighing is some require not only a longitudinal center of gravity but also a lateral center of gravity. If a lateral center of gravity is required it will be provided in the FAA Data Sheet or the FAA Specifications Sheet whichever is applicable to the particular aircraft.

Many of our helicopters today will have varying center of gravity limits depending on the gross weight of the aircraft and the type of work which they are doing. For example, a certain helicopter at 2650 pounds has the CG range from (−3.0) to (+3.5) and at 2100 pounds the range is (−3.0) to (+4.0). Another helicopter might have limitations which change the CG limits as to the type of landing and takeoffs made. For example, a certain helicopter has different limits for vertical and edge procedures.

Below are two weight and balance calculations for helicopters.

PROBLEM 1 H

The first helicopter to be worked is a Bell 47 D1. From the aircraft specifications data sheet the following information is obtained:

Datum: Station (0): Centerline of weld cluster just forward of the leveling lugs. (Weld cluster is approximately 2 inches forward of the centerline of the mast.)

Leveling Means: Leveling lugs lower left hand longeron aft of mast and adjacent cross tube.

CG Range: (−2.0) to (+2.9).

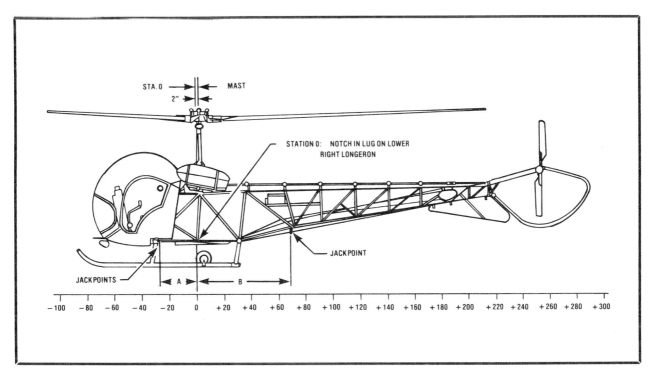

Fig. 3-25   View of Bell 47.

Empty CG Range: Graph

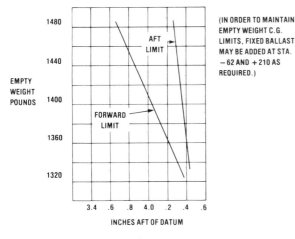

(IN ORDER TO MAINTAIN
EMPTY WEIGHT C.G.
LIMITS, FIXED BALLAST
MAY BE ADDED AT STA.
−62 AND +210 AS
REQUIRED.)

Fig.   3-26

Fuel Capacity: 29 gallons (+24).

Oil Capacity: 3 gallons (+5) (including 1 gallon unusable).

Maximum Weight: 2350 pounds with (200 HP).

Number of Seats: 3 (Pilot and 2 passengers) (−30).

From the manufacturer's manual the following information is obtained:

Weighing Points: Jack points, two forward of the datum, one aft.

Rotor Position: Fore and aft (no tiedown).

Battery Location: Fore (−67) aft (+96).

View of Bell 47D1:   Refer to Fig. 3-25.

Step 1:   Clean the helicopter and remove all loose articles from the aircraft and drain the fuel and oil.

Step 2:   Place the helicopter in a hangar with a fairly level floor. Close the hangar doors and untie the main rotor.

Step 3:   Check the scale adjustment for zero and place one jack on each scale under each jackpoint.

Step 4:   Place a spirit level on the leveling lugs and raise the helicopter off the floor. Check for level both longitudinally and laterally.

Step 5:   Record the scale readings and subtract tare (jacks).

41

	Gross	Tare	Net
Left Front	530	11	519
Right Front	530	11	519
Tail	520	7	513
Total Empty Weight:			1551 pounds

Step 6: Using a plumb bob and a steel tape, measure the horizontal distance from the front jackpoint to the rear jackpoint and from the front jackpoint to the datum. (The datum is marked by a notch on the lower right longeron of the center frame.) The distances are: (+96) and (−28).

Step 7: Using the formula for tail wheel aircraft compute the empty CG as discussed previously:

$$CG = -D + \frac{R \times L}{W}$$

$$-28 + \frac{513 \times 96}{1551} = 3.75$$

Our empty CG is located at 3.75 inches aft of the datum.

Step 8: Use the following formula:

CG X Weight = Moment
3.75 X 1551 = 5816.25

Step 9: The useful load of the helicopter is the empty weight subtracted from the maximum weight or 2350 − 1551 = 799 pounds. Useful load = 799 pounds.

PROBLEM 2 H

For the second helicopter problem we will use a Bell 206B. From the Aircraft Data Sheet the following information is obtained:

Datum: 1 inch forward of the most forward point of the fuselage nose section (55.16 inches forward of the jackpoint center line.)

Leveling Means: Plumb line from ceiling left rear cabin to index plate on floor. Serial No. 584 and subsequent.

Longitudinal CG Range: (+106) to (+111.4) at 3200 pounds; (+106) to (+112.1) at 3000 pounds; (+106) to (+112.4) at 2900 pounds; (+106) to (+113.4) at 2600 pounds; (+106) to (+114.2) at 2350 pounds.

Lateral Vs. Longitudinal CG Limits: See Fig. 3-27

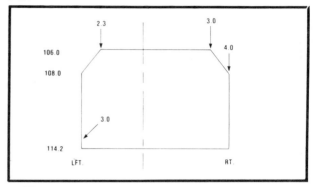

Fig. 3-27   Lateral vs. Longitudinal CG limits.

Empty CG Range: (+114.9) to (+116.6) at 1850 pounds; (+115.8) to (+117.6) at 1690 pounds; (+117.0) to (+118.0) at 1500 pounds; (+118.6) at 1325 pounds.

When Empty Weight CG falls within the range given, complete computations of critical fore and aft CG positions are unnecessary. Refer to Fig. 3-28

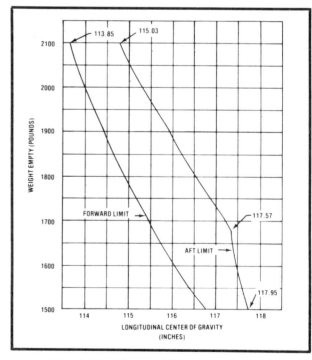

Fig. 3-28   Empty CG limits graph.

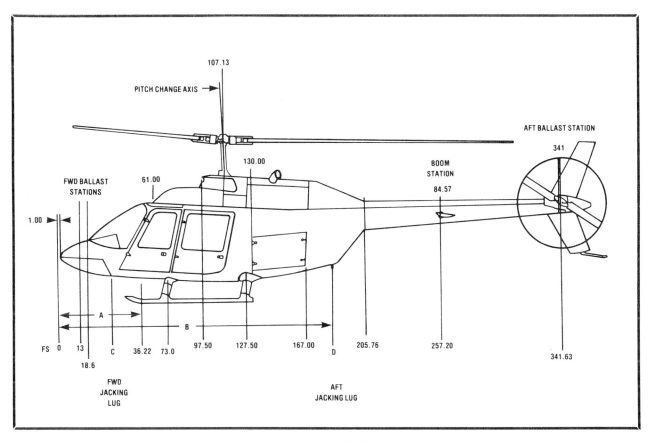

Fig. 3-29   View of Bell 206B.

Maximum Weight: 3200 pounds.

Minimum Crew: 1 at (+65).

Passengers: 1 at (+65) and 3 at (+104).

Maximum Cargo: 1200 pounds.

Fuel Capacity: 76 gallons (+116.1) unusable fuel 10 pounds at (120).

Oil Capacity: 5.5 quarts (+179.0) unusable oil, 2 quarts undrainable oil - 1.0 pound at (+167.0).

From the Manufacturer's Manual the following information is obtained:

Weighing Points: Jackpads 2 at (+55.16); 2 at (+179.92).

View of 206B: Refer to Fig. 3-29.

Step 1:   Clean the helicopter and remove all loose articles from the aircraft. (Check baggage compartment).

Step 2:   Drain the fuel and oil from the engine. Check the hydraulic and transmission reservoir and fill to the required level. Add ten pounds of fuel to the empty tank. NOTE: This could be done mathmatically as we did with the 172M.

Step 3:   Place the helicopter in a hangar with a level flow. Close the hangar doors and remove the main rotor tiedown.

Step 4:   Check the scale adjustment for zero and place one jack on the center of each scale under each jack point.

Step 5:   Hang plumb bob from the slotted plate in the cabin roof directly above the leveling plate and level the aircraft both longitudinally and laterally.

43

Step 6: Record the scale readings and subtract the tare.

	Gross	Tare	Net
Left Front	407.5	25.0	382.5
Right Front	405.0	25.0	380.0
Aft	873.7	79.0	794.7
			1557.2

Total Weight: 1557.2 pounds

Step 7: Using the manufacturer's information obtain the horizontal distance from the front jackpoint to the rear jackpoint and the front jackpoint to the datum. The distances are (124.76) and (55.16) respectively.

Step 8: Using the formula for tail wheel aircraft, compute the empty CG:

$$CG = D + \frac{R \times L}{W}$$

$$55.16 + \frac{794.7 \times 124.76}{1557.2} = 118.83$$

Step 9: Use the following formula to determine the empty moment:

CG X Weight = Moment
118.83 X 1557.2 = 185042.1
Empty Moment = 185042.1

Step 10: The useful load of the helicopter is the empty weight subtracted from the maximum weight or 3200 − 1557.2 = 1642.8 pounds. Useful load is 1642.8 pounds.

The calculation of the lateral center of gravity is quite similar to the longitudinal center of gravity except that the Butt Line "0" or center line is used as the datum. Butt lines to the right of the center line are designated as plus. The forward net scale readings are multiplied by the inches outboard of the center line to obtain the moments

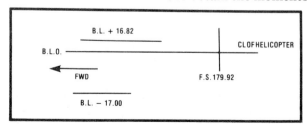

Fig. 3-30 Lateral CG drawing.

with the aft scale on the center line. Using the total moment divided by the total weight formula, calculate the lateral CG. Refer to Fig. 3-21.

	Scale	Tare	Net
Left Forward Jackpoint	407.5	25.0	382.5
Right Forward Jackpoint	405.0	25.0	380.0
Aft Jackpoint	873.7	79.0	794.7
	1686.2	129.0	1557.2

$$\text{Lateral CG} = \frac{(-17.00(382.5)(+16.82)(380)}{1557.2}$$

$$\text{Lateral CG} = -0.07 \text{ inches.}$$

QUESTIONS:

Problem: Cessna 310 L   Serial No. 310L-0100

The following information is obtained from the Aircraft Specifications:

Datum: Forward face of fuselage bulkhead forward of rudder pedals.

Leveling Means: External splice plate on left side of fuselage under the windows.

CG Range: (+38) to (+43.1) at 5200 pounds; (+43.6) at 4800 pounds; (+32.0) to (+43.6) at 4300 pounds or less.

Number of Seats: 5 (standard) (2 at (+37) and 3 at (+71))

Fuel Capacity: 143 2 wingtip tanks 51 gallon each at (+35) and 2 auxiliary tanks 20.5 gallons each at (+47). See Note 1.

Oil Capacity: 24 quart (12 quart each at (−3.5). 6 quart unusable per engine. See Note 1.

Note 1: Certified empty weight and corresponding center of gravity location must include undrainable oil (not included in the oil capacity) and unusable fuel (not included in the fuel capacity) as follows:

Fuel:	12 pounds at (+44)
Oil:	1 pound at (−3.5)

View of Cessna 310L: Refer to Fig. 3-31.

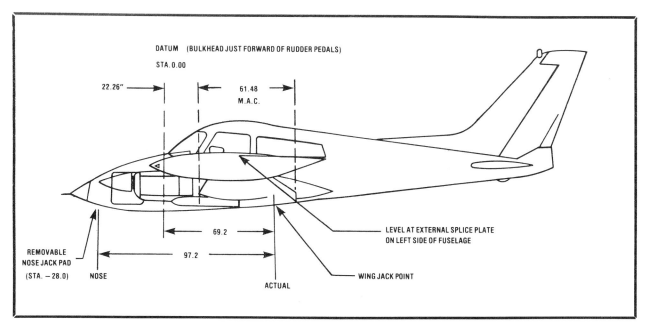

DATUM (BULKHEAD JUST FORWARD OF RUDDER PEDALS)

STA. 0.00

22.26"

61.48
M.A.C.

69.2

97.2

REMOVABLE
NOSE JACK PAD
(STA. −28.0) NOSE

ACTUAL

LEVEL AT EXTERNAL SPLICE PLATE
ON LEFT SIDE OF FUSELAGE

WING JACK POINT

Fig. 3-31   Cessna 310L

Calculate the following:

Given	Scale	Tare	Net
Left Wheel	1144	0	?
Right Wheel	1074	0	?
Nose Wheel	1204	16	?
Total Empty Weight:			?

Note: Fuel and Oil Drained.

Maximum Weight: 5200 lbs.

Measurements: Datum to Main Gear: 69.2; Main Gear to Nose Gear: 97.2.

Calculate empty CG: Total Empty Moment; Useful Load.

QUESTIONS:

1. In the problem, what is to be done with the unusable fuel?

    A.   Subtract from the Empty Weight.

    B.   Add to the Empty Weight.

    C.   Add to the Empty CG and Weight.

    D.   Nothing, undrainable is the same as unusuable.

2. What is to be done with the unusable oil?

    A.   Subtract from the Empty Weight.

    B.   Add to the Empty Weight.

    C.   Add to the Empty CG.

    D.   Nothing.

3. Undrainable oil is the same as unusable.

    A.   True

    B.   False

4. The Total Empty Weight of the aircraft is:

    A.   5200 pounds

    B.   3419 pounds

    C.   3406 pounds

5. The formula for locating the Empty CG is:

    A. $CG = D - \dfrac{F \times L}{W}$

    B. $CG = -D + \dfrac{F \times L}{W}$

45

C.
$$CG = D + \frac{R \times L}{W}$$

6. What is the Empty Certificated CG?

   A. 35.3

   B. 42.4

   C. 35.6

7. What is the Useful Load?

   A. 5200

   B. 1782

   C. 1794

8. What is the Total Certificated Empty Moment?

   A. 120747.6

   B. 120231.8

   C. 1444414.4

9. The aircraft is within the Empty CG Range:

   A. True

   B. False

   C. The Range is not given.

10. In reference to Problem 1H was the helicopter within Empty CG limits?

    A. Yes

    B. No

11. In reference to Problem 2H was the helicopter within lateral CG limits?

    A. Yes

    B. No

12. The PA24-250 may be operated in dual categories.

    A. Yes

    B. No

13. In Problem 1T the Normal Category Maximum Weight was higher than the Restricted Category.

    A. True

    B. False

14. What is the formula for adding the unusable fuel in Problem 1 N?

    A. $\dfrac{TW}{TM} = CG$

    B. $\dfrac{TM}{TW} = CG$

    C. $M = A \times W$

15. The CG may be expressed as a distance from LEMAC to TEMAC.

    A. True.

    B. False.

CHAPTER IV
Aircraft Loading

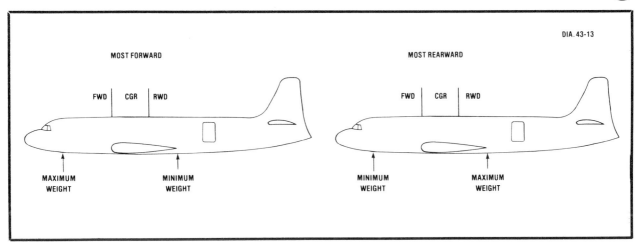

Fig. 4-1 Most forward and most rearward center of gravity investigation.

The "Empty Weight" and "Empty Center of Gravity" are the basis for *all* of the calculations of loading the aircraft and any further calculations that may be made.

Most aircraft manufactured today make use of loading charts, tables, computers, or placards for all normal loading of the aircraft based on the empty weight and the center of gravity figures. This responsibility for the proper loading is that of the operator of the aircraft. However, there are many aircraft still flying today that have no provisions for loading; this means the Empty Center of Gravity and Weight are rather meaningless figures while the Loaded Center of Gravity and Maximum Weight are necessary to maintain safe flight conditions.

Usually, if an Empty Center of Gravity is given for the particular aircraft and the aircraft falls within this range, no further calculations will be necessary (provided that the aircraft is loaded in accordance with the manufacturer's recommendations). These recommendations may be in the form of placards or may be contained in the approved flight manual.

A. Most Forward and Rearward Investigations

If the aircraft has undergone modification or equipment changes, and no empty center of gravity range is given, then the loaded extreme center of gravity should be investigated. These extremes would be the most forward and most rearward conditions.

In some instances, it might also be advisable to calculate the loading of aircraft in different configurations for ease in loading computations. This is often done when the aircraft is being used for the same purposes continuously where no more than three or four different configurations would be used.

The investigation of the extreme conditions forward and rearward will be based on the empty center of gravity calculations of the aircraft.

The forward condition is investigated by simply loading the aircraft in such a manner that all useful load forward of the loaded center of gravity range is at a maximum weight and all minimum weights necessary for flight are used aft of the forward limit.

The rearward condition is investigated in a similar manner. All useful load aft of the rearward loaded center of gravity range is at maximum weight and minimum weights necessary for flight are used forward of the aft limit. Refer to Fig. 4-1.

47

For all calculations dealing with extreme conditions, the following standards of weight will be used:

Gasoline: 6 pounds per U.S. Gallon
Turbine Fuel: 6.7 pounds per U.S. Gallon
Lubricating Oil: 7.5 pounds per U.S. Gallon
Crew and Passengers: 170 pounds

The one item which is continuously changing weight is fuel. For this reason, in the most forward and rearward calculations the term *minimum fuel* (METO FUEL) will be used. This is the minimum fuel for calculation purposes of the most forward and rearward conditions. It is calculated by using the formula of 1/12 of a gallon for each horsepower at maximum except takeoff horsepower.

If the METO horsepower for a particular aircraft was 240 horsepower, then the minimum fuel would be $\frac{240}{12}$ or 20 gallons. To convert gallons to pounds simply multiply by 6 or 6 X 20 = 120 pounds of gasoline. Usually in order to simplify the mathmatics, the METO horsepower is divided by 2 or $\frac{240}{2}$ = 120 pound of gasoline for minimum fuel.

It must be noted, however, that this formula is only applicable to reciprocating engines. Turbine engines have such a variation in "Specific Fuel Consumption" that no set formulas are possible for this calculation. The minimum fuel information is furnished by the manufacturer.

This minimum fuel will be used when the fuel tank lies behind the forward limit when calculating the most forward center of gravity, and ahead of the aft limit when figuring the most rearward center of gravity.

If the fuel tank is ahead of the forward limit, full fuel should be calculated at the standard weight.

If the fuel tank is behind the most rearward center of gravity, then the fuel tank should be full when the most rearward center of gravity is calculated.

Oil, being a necessary item for flight, will be included at full capacity in the most forward and most rearward center of gravity calculation. Some aircraft have a basic weight and with these aircraft the oil is already included in the aircraft weight.

Pilot and passengers are all given the standard weight of 170 pounds. The pilot is necessary for flight and must be included in all calculations. In some of the older aircraft a tandem seating arrangement is used. This tandem seating arrangement often had controls at both seats and were placarded as to which seat is used for solo flight because of the center of gravity limits. This seat location must be considered when calculating the forward and rearward center of gravity when it is being investigated.

The passengers during the investigation of forward and rearward limits will be moved in relation to the seats position to the most forward and rearward limit.

The baggage, like the passengers, will be placed in the aircraft only if the location of the baggage is ahead of the forward limit in the calculations of the most forward center of gravity, and behind the aft limit for the most rearward center of gravity. In either of these instances, the maximum baggage weight will be used in the investigation.

The normal procedure for calculating the most forward and rearward center of gravity is to list the items to be included in the useful load by weight, arm, and moment in columns to include the empty aircraft weight, arm, and moment. The weight column is added giving the total weight of the aircraft. The individual weights are multiplied by the arm giving the individual moments. The moment column is then added giving the total moments. Using the Total Moment divided by the Total Weight Formula the new center of gravity may be located.

A typical aircraft that will require an investigation of the most forward and rearward conditions would be the same PA24-250 that the empty center of gravity was calculated for in Problem 2N. Although the empty center of gravity and weight were calculated, there are no provisions made to figure the loaded center of gravity, and no empty center of gravity was given. No provisions are available for determining the loaded center of gravity except than by sample loadings or by actual calculation.

From the Aircraft Specification Sheet, it was determined that the center of gravity range, seat, baggage, fuel, oil locations and weights were as follows:

CG
Range: (+86.0) to (+93.0) at 2900 lbs.
(+82.5) to (+93.0) at 2600 lbs.
(+80.5) to (+93.0) at 2000 lbs.

Maximum Weight: 2900 pounds.

Number of Seats: 4 -- 2 at (+85) and 2 at (+118.5).

Maximum Baggage: 200 pounds (+142).

Fuel Capacity: 56 gallons (two 28 gallon tanks) (+90).

Oil Capacity: 3 gallons (+28).

1. Most forward investigation

Step 1: The first item necessary is the Empty Aircraft. This information will be obtained from Problem 2 N.

	Weight	Arm	Moment
Empty Aircraft	1711	82.0	140302

Step 2: The first item of Useful Load will be fuel. The arm of the fuel is (+90) and the forward center of gravity limit is a variable figure with the most aft position being (+86). This means that minimum fuel will be used. METO divided by two, is the formula for minimum fuel by weight. The horsepower is 250 Maximum Except Takeoff.

$$\frac{250}{2} = 125 \text{ pounds of fuel.}$$

The moment of the fuel is the Arm times the Weight or 125 X +90 = 11250 Moment. The fuel will be added to the column with the empty weight of the aircraft.

	Weight	Arm	Moment
Empty Aircraft	1711	+82	140302
Fuel	125	+90	11250

Step 3: Oil is necessary for flight so full oil must be added. From the Specification Sheet, the oil capacity is 3 gallons. The standard weight for oil is 7.5 pounds per gallon. By multiplying the gallons by the weight per gallon, the total weight is obtained or 7.5 X 3 = 22.5. The length of the arm is (+28). The moment of the oil is 22.5 X 28 or 630. This will also be added to the column with the Aircraft and the Fuel.

	Weight	Arm	Moment
Empty Aircraft	1711.0	+82	140302
Fuel	125.0	+90	11250
Oil	22.5	+28	630

Step 4: From the Specification Sheet the seat locations are obtained. They are at +85 and +118.5. It is quite easy to determine that no passengers will be placed in the rear seat. However, the forward range is variable with the weight of the aircraft. In all cases except Maximum Weight, the front seat is behind the forward limit. With minimum fuel and no passengers in the rear seats, it will not be possible to reach Maximum Weight so it will not be necessary to include both front seats in the investigation. The pilot is necessary for flight and he will be placed at (+85) using the standard weight of 170 pounds. The moment, using the Weight X Arm Formula will be 170 X 85 or 14450. These figures will again be added to the columns of the empty aircraft, fuel and oil.

	Weight	Arm	Moment
Empty Aircraft	1711.0	+82	140302
Fuel	125.0	+90	11250
Oil	22.5	+28	630
Pilot	170.0	+85	14450

Step 5: The baggage will not be necessary for flight and is located behind the forward limit. Therefore no baggage will be included in the investigation.

Step 6: The formula for the center of gravity is Total Moment Divided by Total Weight, so the columns of weight and moment will be added up and divided.

	Weight	Arm	Moment
Empty Aircraft	1711.0	+82	140302
Fuel	125,0	+90	11250
Oil	22.5	+28	630
Pilot	170.0	+85	14450
	2028.5		166632

$$CG = \frac{TM}{TW} \text{ or } \frac{166632}{2028.5} = 82.1$$

Most forward is 82.1. The most forward limit is not exceeded, so no further action will be necessary.

2. *Most rearward investigation*

Step 1: The first item necessary will be the Empty Aircraft. This information is obtained from Problem 2 N.

	Weight	Arm	Moment
Empty Aircraft	1711	+82	140302

Step 2: The fuel is located at (+90) which is forward of the rearward center of gravity. Therefore minimum fuel will be used. Since minimum fuel was calculated for the most forward investigation it will not be necessary to recompute for the rearward investigation. It will simply be added to the column with the empty weight of the aircraft.

	Weight	Arm	Moment
Empty Aircraft	1711	+82	140302
Fuel	125	+90	11250

Step 3: The oil is necessary for flight and has already been calculated in the most forward center of gravity investigation, so the oil is placed in the column with the Empty Aircraft and the Fuel.

	Weight	Arm	Moment
Empty Aircraft	1711.0	+82	140302
Fuel	125.0	+90	11250
Oil	22.5	+28	630

Step 4: The pilot will be necessary for flight. The other front seat is ahead of the rearward limit so it will not be necessary to fill the seat. The two rear seats are behind the rearward center of gravity limits at (+118.5). Both of these seats must be filled with standard weight passengers and the moment calculated. The pilot and two passengers in the rear seats will be added to the weight, arm, and moment columns.

	Weight	Arm	Moment
Empty Aircraft	1711.0	+82.0	140302
Fuel	125.0	+90.0	11250
Oil	22.5	+28.0	630
Pilot	170.0	+85.5	14450
Passengers (2)	340.0	+118.5	40290

Step 5: The baggage compartment is located at (+142) and has a maximum weight of 200 pounds. This is behind the rearward center of gravity limit and must be included in the most rearward investigations at maximum capacity. The weight, arm and moment will be added to the columns.

	Weight	Arm	Moment
Empty Aircraft	1711.0	+82.0	140302
Fuel	125.0	+90.0	11250
Oil	22.5	+28.0	630
Passengers	340.0	+118.5	40290
Pilot	170.0	+85.0	14450
Baggage	200.0	+142.0	28400

Step 6: The total weight and moment columns are added giving a total weight of 2568.5 pounds and a total moment of 235322. Using the total moment divided by total weight formula, the most rearward center of gravity is located at 91.6 inches, which is within the center of gravity limits.

In summary, the most forward and most rearward investigations of this revealed no adverse conditions in the center of gravity location. This aircraft also had no negative arms leaving negative moments which simplified the investigation.

B. Sample Loadings

In order to simplify the loading of the aircraft of the type that has no graph or other means of calculation, the loaded center of gravity sample loadings are often carried in the aircraft. This information is often provided especially if the aircraft is operated for air taxi and the loadings can be computed by the maintenance personnel or the pilot.

The same method may be used for these sample loading configurations as is used for the investigation of the most forward and most rearward center of gravity. Since many of the weight, arms, and moments have already been calculated for the PA24-250, the same aircraft will be used in calculating sample loadings.

It will make no difference which sample loading will be computed first but it is normally an accepted practice to start with forward loading and work aft since much of the forward information will also be used for rearward computations.

1. Forward loading

Since the most forward center of gravity investigation included one occupant, the first sample loading will be two occupants and full fuel.

Step 1: The empty weight of the aircraft was computed in Problem 2 N. The oil was computed in the most forward center of gravity investigation so it will not be necessary to repeat the computation. The fuel capacity is 2 tanks (27 gallons each at (+90)). Using the standard weight of 6 pounds or a total of 324 pounds this will be carried with a moment of 29160.

Step 2: Two occupants (pilot and passenger) are placed in the front seat which is located at +85. Using the standard weight of 170 pounds or a total of 340 with a moment of 28900.

Step 3: Placing the empty weight of the aircraft, full fuel, oil, and one pilot and passenger in columns of weight, arm, and moments, the total weight and moment are

obtained. Using the Total Weight divided by Total Moment Formula the loaded center of gravity may be computed.

	Weight	Arm	Moment
Empty Aircraft	1711.0	+82	140302
Oil	22.5	+28	630
Fuel	324.0	+90	29160
Pilot and Passenger	340.0	+85	28900
	2397.5		198992

$$CG = \frac{198992.0}{2397.5} = 83 \text{ inches}$$

2. Maximum loading

The same method can be used for a maximum load to determine if the maximum weight and center of gravity would be exceeded. This loading would include full fuel, oil, seats and baggage as follows:

	Weight	Arm	Moment
Empty Aircraft	1711.0	+82.0	140302
Fuel	324.0	+90.0	29160
Oil	22.5	+28.0	630
Pilot & Passengers	340.0	+85.0	28900
2 Passengers	340.0	+118.5	40290
Baggage	200.0	+142.0	28400
	2937.5		267682

New CG = +91.1

The maximum weight has been exceeded in this problem by 37.5 pounds. The center of gravity, however, is within limits. In this type of situation, something must be removed from the aircraft load. In most instances fuel would be limited but baggage could also be limited.

C. The Graph Method

Another system used today is the *graph method* which uses two graphs to determine the useful load for the aircraft. An aircraft utilizing such a system is the 172 M used in Problem 1N (Fig. 4-2).

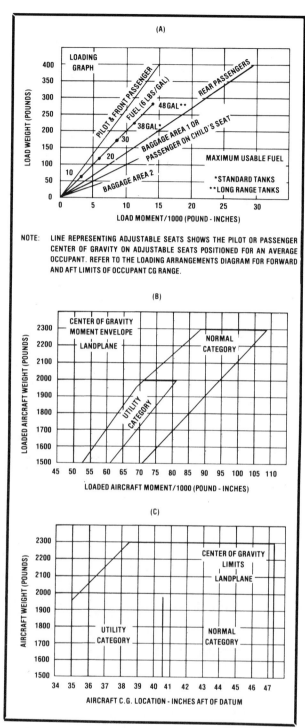

Fig. 4-2   Graph method of loading for 172M.

In viewing Fig. 4-2A, all items of useful load are represented by diagonal lines. (The only item not shown in this manner is the oil. In Problem 1N, the oil was included in the Basic Weight.) On the left side of the graph the weight is represented in pounds. At the bottom of the graph the load moment is shown in 1000 (pound-inches). To deter-

mine the moment of any item, the weight is entered on the left side of the graph to the point that it intersects the diagonal line representing the item. At the point of intersection a straight line is followed to the bottom of the graph where the item moment is found. This step will be repeated for each item included in the load desired to be carried.

These items will be listed and the weight and moment totaled with the empty aircraft weight and moment.

The graph in Fig. 4-2B represents the operating center of gravity range by the enclosed diagonal lines for both the normal and utility categories of the aircraft. The weight is indicated in pounds on the left side and the moment is given in thousands across the bottom.

The total weight and moment from the desired aircraft loading are entered into the graph. If the weight and moment intersect within the envelope, the aircraft is within the loaded center of gravity range.

EXAMPLE 1

The empty center of gravity was determined to be 39.7 inches with a basic weight of 1369 pounds and a basic empty moment of 54500.5.

Step 1:   The weight of the aircraft will remain the same but the moment must be reduced by 1000 to be utilized with the graphs. The moment will now read 54.5 rather than 54500.5. This is done for purposes of simplification.

Step 2:   Determine the load to be carried by the aircraft. For our problem, three passengers and the pilot will be carried. Full fuel with standard tanks and 20 pounds of baggage in area 1 will also be used.

   (a)   The pilot and passengers in the front seats weigh 170 pounds each making a total of 340 pounds. Entering this weight in the left-hand weight column, it will intersect the pilot passenger line at 12.6 moment.

   (b)   The full standard tanks hold 38 gallons. This amount of fuel is marked on the diagonal line for the fuel so that there is no need to determine the fuel weight. By fol-

52

lowing the gallons of fuel to the left, the weight is 228 pounds. By following the Fuel Quantity down, the moment is 10.9.

(c) Two rear passengers also have a total weight of 340 pounds. By entering this weight on the weight column to intersect the diagonal line, the moment is 24.8.

(d) The baggage at 20 pounds in the graph has a moment of 2.

Step 3: Place the Basic Aircraft Weight and Moment in a column with the Intended Useful Load and add the weights and moments.

Basic Aircraft	1369	54.5
Pilot and Passengers	340	12.6
Fuel	228	10.9
Rear Seat Passengers	340	24.8
Baggage	20	2.0
	2297	104.8

Step 4: Place the weight and moments into Chart B to determine if the load is within the center of gravity envelope.

In summary, it can be seen that by adding additional weight in the form of baggage to areas 1 and 2, the aircraft can exceed the maximum weight as well as be placed out of the center of gravity envelope.

EXAMPLE 2

Another aircraft that utilizes graphs is the Ag Truck used in Problem 1T of this text. This aircraft utilizes the same type of graph as the 172 M except that the envelope is of a different shape and a separate graph is used for the density of the various liquids used (Fig. 4-3).

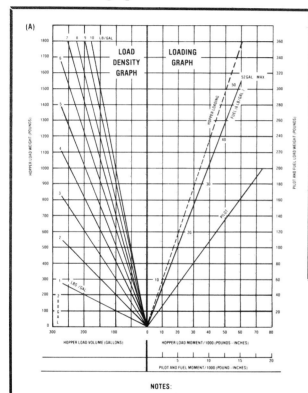

NOTES:

(1) Line representing pilot shows the pilot center of gravity on adjustable seat positioned for an average occupant. Refer to the Sample Loading Problem for forward and aft limits of occupant C.G. range.

(2) The hopper loading line is based on a density of 8.345 Lbs./U.S. Gallons. This line may be used with reasonable accuracy for materials with a density other than 8.345 Lbs./U.S. Gallons.

(3) Conversion factors are given so that the density of the material being applied may be converted into lbs/gallon which can be used in conjunction with the Load Density Graph and Hopper Loading Graph. Conversion factors are as follows: Lbs/U.S. Bushel x .1074 = Lbs/U.S. Gallon
Lbs/Cubic Feet x .1337 = Lbs/U.S. Gallon

(4) TO CALCULATE THE WEIGHT OF A HOPPER LOAD, locate the intended volume to be carried at the bottom of the Load Density Graph and read upward until intersecting the material density line (1 thru 10 lbs/gallon) for the material being used; then read across to the left to find the weight of the hopper load. Check to see that this weight does not cause the gross weight of the aircraft to be exceeded.

(5) TO CALCULATE THE VOLUME OF THE DESIRED HOPPER LOAD, locate the weight of the hopper load on the left side of the Load Density Graph and read across until intersecting the material density line (1 thru 10 lbs/gallon) for the material being used; then read down to the bottom to find the volume of the hopper load in gallons. If the volume exceeds the volume capacity of the hopper, reduce the weight accordingly.

(6) TO CALCULATE THE MOMENT/1000 FOR HOPPER LOADS, locate the weight of the hopper load (as determined in steps 4 or 5 above) on the left side of the Hopper Loading Graph and read across to the right until intersecting the line representing the aircraft hopper. From this point, drop down vertically and read the moment/1000. Write the weight and moment/1000 in the Sample Loading Problem.

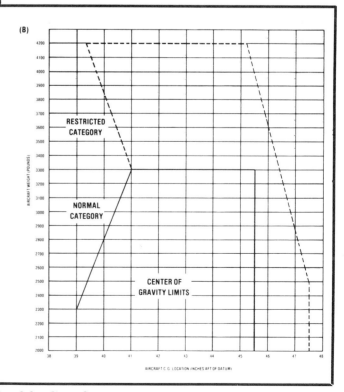

Fig. 4-3 Ag Truck loading charts.

Step 1: The Basic Weight in Problem 1T was determined to be 2450 and the moment was 96162.5. The moment is divided by 1000 leaving the moment at 96.2 (rounded off to the nearest number).

Step 2: Using Graph A, the pilot weighs 170 pounds. His moment is 15.8. The maximum fuel load is 52 gallons which is 312 pounds with a moment of 15.

Step 3: Place the Basic Weight and Moment of the aircraft and the Useful Load in columns and add these items.

	Weight	Moment
Basic Aircraft	2450	96.2
Fuel	312	15.0
Pilot	170	15.8
	2932	127.0

Step 4: Place the Total Weight and Moment into Graph B to determine if the aircraft is within the center of gravity envelope. It is within the acceptable limits with the hopper empty.

Step 5: Place 250 gallons of liquid in the hopper that weighs 3 pounds per gallon. This will weigh 750 pounds using the left-hand side of Graph A. Follow the hopper weight line from the left-hand side of Graph A to its intersection point of the diagonal hopper line. Follow the line down from the intersection to find the moment. The moment is 32.5

Step 6: Add the hopper weight and moment to the total weight and moment of the aircraft, pilot and fuel.

	Weight	Moment
Aircraft, Pilot, and Fuel	2932	127.0
Hopper	750	32.5
	3682	159.5

Step 7: Follow the weight and moment lines to determine if the aircraft is within the center of gravity envelope. If this aircraft was filled to the hopper capacity with a dense material, the maximum weight and center of gravity would have been exceeded.

1. The table-graph method

Another method used on light aircraft today is the table-graph method as used on the Model 36 Beech Bonanza. This method makes use of tables in order to obtain the weight, arm, and moment of items of the useful load rather than the graph used on the 172 M.

After the Useful Load and the Empty Aircraft are totaled, these figures are used in a weight-moment graph in order to determine the center of gravity location in the envelope.

The aircraft, used as an example, will have a maximum weight of 3600 pounds and an empty weight of 2082 pounds. The moment will be 160300. This moment will be divided by 100 in order to reduce the size of the figure so that the index figure will be 1603 for the Empty Weight Moment (Fig. 4-4).

(A)

	OCCUPANTS		
WEIGHT	FRONT SEATS	CENTER SEATS	AFT SEATS
	ARM 75	ARM 115	ARM 146
		MOMENT/100	
100	75	115	146
110	83	127	161
120	90	138	175
130	98	150	190
140	105	161	204
150	113	173	219
160	120	184	234
170	128	196	248
180	135	207	263
190	143	219	277
200	150	230	292

(B)

FUEL LEADING EDGE TANKS ARM 75		
GALLONS	WEIGHT	MOM/100
5	30	23
10	60	45
15	90	68
20	120	90
25	150	113
30	180	135
35	210	158
40	240	180
45	270	203
49	294	221
55	330	248
60	360	270
65	390	293
70	420	315
75	450	338
80	480	360

Fig. 4-4 Loading tables for Model 36 Beech Bonanza.

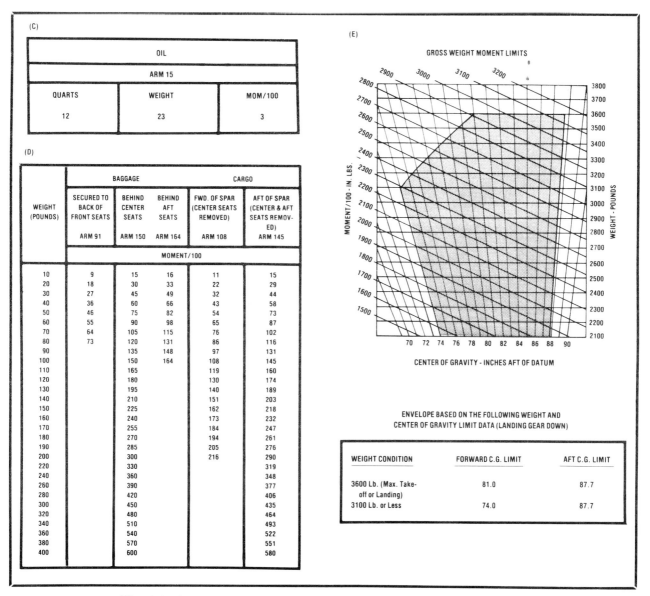

(C)

OIL		
ARM 15		
QUARTS	WEIGHT	MOM/100
12	23	3

(D)

WEIGHT (POUNDS)	BAGGAGE			CARGO	
	SECURED TO BACK OF FRONT SEATS ARM 91	BEHIND CENTER SEATS ARM 150	BEHIND AFT SEATS ARM 164	FWD. OF SPAR (CENTER SEATS REMOVED) ARM 108	AFT OF SPAR (CENTER & AFT SEATS REMOVED) ARM 145
	MOMENT/100				
10	9	15	16	11	15
20	18	30	33	22	29
30	27	45	49	32	44
40	36	60	66	43	58
50	46	75	82	54	73
60	55	90	98	65	87
70	64	105	115	76	102
80	73	120	131	86	116
90		135	148	97	131
100		150	164	108	145
110		165		119	160
120		180		130	174
130		195		140	189
140		210		151	203
150		225		162	218
160		240		173	232
170		255		184	247
180		270		194	261
190		285		205	276
200		300		216	290
220		330			319
240		360			348
260		390			377
280		420			406
300		450			435
320		480			464
340		510			493
360		540			522
380		570			551
400		600			580

(E)

GROSS WEIGHT MOMENT LIMITS

MOMENT/100 - IN. LBS.

WEIGHT - POUNDS

CENTER OF GRAVITY - INCHES AFT OF DATUM

ENVELOPE BASED ON THE FOLLOWING WEIGHT AND CENTER OF GRAVITY LIMIT DATA (LANDING GEAR DOWN)

WEIGHT CONDITION	FORWARD C.G. LIMIT	AFT C.G. LIMIT
3600 Lb. (Max. Take-off or Landing)	81.0	87.7
3100 Lb. or Less	74.0	87.7

Fig. 4-4   (continued)   Loading tables for Model 36 Beech Bonanza.

EXAMPLE 1

Step 1: Record the Empty Weight of the aircraft and the Index Moment. Record the weight and corresponding moment of each useful load item to be carried from the weight and moment tables, and then add the weights and moments.

	Weight	Moment
Empty aircraft	2082	1603
Oil (12 quarts)	23	3
Fuel (80 gallons)	480	360
Front Seats	340	255
Center Seats	340	391
Aft Seats	250	365
Baggage (Secured to Front Seats)	60	55
Baggage (Behind Aft Seats)	25	41
Total	3600	3073

Step 2: Use the total weight and moment in Graph 4-4E to determine if the aircraft is within the approved center of gravity range.

Step 3: Since fuel is being consumed during the flight, the center of gravity location will change. This change could adversely affect the flight characteristics of the aircraft. Calculate the amount of fuel to be consumed. Find the weight and moment from the table and subtract this amount from the Total Weight and Moment at takeoff.

Total Weight			
Takeoff	3600	Total Moment	3073
Fuel Consumed			
(60 gal.)	−360	Fuel Moment	− 270
	3240		2803

Step 4: Check Graph 4-4E to see if the aircraft is still within the center of gravity envelope. If the aircraft is not within the envelope, the load must be rearranged.

2. The table method

A system that is very similar to the system used on the Model 36 Bonanza is the C34R Sierra. The only great difference is the complete use of tables with *no* graph for the loaded center of gravity. What has been done is to simply place various weights and corresponding moments on a chart. The moments represent the maximum fore and aft limits of the center of gravity range of the aircraft at a specific weight.

The aircraft will have a maximum weight of 2750 pounds and a Basic Empty Weight of 1720 pounds. The Basic Empty Weight Moment is 191200. The moment is divided by 100 in order to reduce the figure. So, now the index moment figure for the basic empty weight will be 1912 (Fig. 4-5).

Step 1: Record the Basic Empty Weight of the aircraft and the corresponding index moment. Using Fig. 4-5 tables A thru E record the weight and corresponding index moment of each useful load item.

	Weight	Moment
Basic Empty Aircraft	1720	1912
Front Seats	340	374
3 and 4 Seats	340	482
5 and 6 Seats	130	222
Fuel	192	225
Total	2722	3215

(A)

USEFUL LOAD WEIGHTS AND MOMENTS

USABLE FUEL ARM 117		
GALLONS	WEIGHT	MOMENT/100
5	30	35
10	60	70
15	90	105
20	120	140
22	132	154
25	150	176
27	162	189
30	180	211
32	192	225
35	210	246
37	222	259
40	240	281
45	270	316
50	300	351
52	312	365
57	342	400

(B)

OCCUPANTS

FRONT SEATS ARM 110		3RD & 4TH SEATS ARM 142	
WEIGHT	MOMENT 100	WEIGHT	MOMENT 100
120	132	120	170
130	143	130	185
140	154	140	199
150	165	150	213
160	176	160	227
170	187	170	241
180	198	180	256
190	209	190	270
200	220	200	284

(C)

OCCUPANTS

5TH & 6TH SEATS ARM 171			
WEIGHT	MOMENT 100	WEIGHT	MOMENT 100
80	137	140	239
90	154	150	257
100	171	160	274
110	188	170	291
120	205	180	308
130	222	190	325
		200	342

Fig. 4-5 Loading charts for C34R Sierra.

MOMENT LIMITS vs WEIGHT LIMITS

Weight	Minimum Moment 100	Maximum Moment 100	Weight	Minimum Moment 100	Maximum Moment 100	Weight	Minimum Moment 100	Maximum Moment 100
1700	1870	2011	2100	2310	2484	2500	2775	2958
1710	1881	2023	2110	2321	2496	2510	2788	2969
1720	1892	2035	2120	2332	2508	2520	2801	2981
1730	1903	2047	2130	2343	2520	2530	2814	2993
1740	1914	2058	2140	2354	2532	2540	2828	3005
1750	1925	2070	2150	2365	2543	2550	2841	3017
1760	1936	2082	2160	2376	2555	2560	2854	3028
1770	1947	2094	2170	2387	2567	2570	2867	3040
1780	1958	2106	2180	2398	2579	2580	2880	3052
1790	1969	2118	2190	2409	2591	2590	2894	3064
1800	1980	2129	2200	2420	2603	2600	2907	3076
1810	1991	2141	2210	2431	2614	2610	2920	3088
1820	2002	2153	2220	2442	2626	2620	2933	3099
1830	2013	2165	2230	2453	2638	2630	2947	3111
1840	2024	2177	2240	2464	2650	2640	2960	3123
1850	2035	2189	2250	2475	2662	2650	2973	3135
1860	2046	2200	2260	2486	2674	2660	2987	3147
1870	2057	2212	2270	2497	2685	2670	3000	3159
1880	2068	2224	2280	2508	2697	2680	3013	3170
1890	2079	2236	2290	2519	2709	2690	3027	3182
1900	2090	2248	2300	2530	2721	2700	3040	3194
1910	2101	2260	2310	2541	2733	2710	3054	3206
1920	2112	2271	2320	2552	2745	2720	3067	3218
1930	2123	2283	2330	2563	2756	2730	3081	3230
1940	2134	2295	2340	2574	2768	2740	3094	3241
1950	2145	2307	2350	2585	2780	2750	3108	3253
1960	2156	2319	2360	2596	2792			
1970	2167	2331	2370	2607	2804			
1980	2178	2342	2380	2619	2815			
1990	2189	2354	2390	2632	2827			
2000	2200	2366	2400	2645	2839			
2010	2211	2378	2410	2658	2851			
2020	2222	2390	2420	2671	2863			
2030	2233	2401	2430	2684	2875			
2040	2244	2413	2440	2697	2887			
2050	2255	2425	2450	2710	2898			
2060	2266	2437	2460	2723	2910			
2070	2277	2449	2470	2736	2922			
2080	2288	2461	2480	2749	2934			
2090	2299	2472	2490	2762	2946			

THE ABOVE WEIGHT AND MOMENT LIMITS ARE BASED ON THE FOLLOWING WEIGHT AND CENTER OF GRAVITY LIMIT DATA:

NORMAL CATEGORY

WEIGHT CONDITION	FWD CG LIMIT	AFT CG LIMIT
2750 (MAX. TAKEOFF OR LANDING)	113.0	118.3
2375 LBS OR LESS	110.0	118.3

(D)

BAGGAGE ARM 167

WEIGHT	MOMENT 100	WEIGHT	MOMENT 100
10	17	140	234
20	33	150	251
30	50	160	267
40	67	170	284
50	84	180	301
60	100	190	317
70	117	200	334
80	134	210	351
90	150	220	367
100	167	230	384
110	184	240	401
120	200	250	418
130	217	260	434
		270	451

Fig. 4-5   (continued)   Loading charts for C34R Sierra.

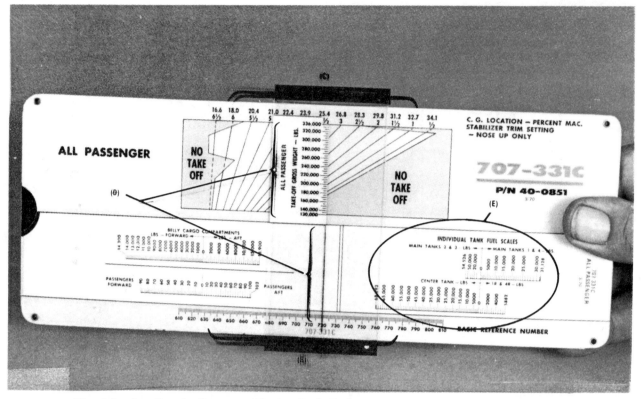

Fig. 4-6 A — Load adjuster   Fig. 4-6 B, C, D, E — Explanations of the load adjuster

Step 2: After obtaining the total moment and weight, go to Table E to see if the aircraft is within weight limitations and center of gravity range.

Step 3: Since fuel is consumed during flight, calculate the amount of fuel to be used. Use the fuel table to determine the weight and moment of this fuel and subtract it from the total weight and moment.

	Weight	Moment
Total	2722	3215
Fuel (25 gal)	−150	−176
	2572	3039

Step 4: Check the landing weight and moment with Table E to see if the aircraft is still within the center of gravity limits.

D. The Load Adjuster Method

Another method used in the calculation of the loaded center of gravity range is the *Load Computer*. This system was first used during World

War II and is still in use today on some aircraft. The system makes use of a slide rule type device in either a conventional slide rule form or in a circular slide rule form. The one that will be used in this text will be of the conventional slide rule form with a moveable cursor and an interscale portion called a slider (Fig. 4-6 A).

At the bottom of the base portion is a basic reference number scale (Fig. 4-6B). This basic reference number scale is derived from the weight and moment of the aircraft.

Since each aircraft type will use a different reference number system, no explanation will be made as to how the basic reference number is obtained. Each aircraft of a particular type will have its own basic reference number depending upon its individual center of gravity location and moment. For example, one may have the basic reference of 690 while another may have its basic reference number as 700.

At the top of the base portion is a scale giving the center of gravity in per cent of MAC and the nose up trim setting for the aircraft (Fig. 4-6C).

The cursor has one hairline and a takeoff weight scale in pounds (Fig. 4-6D).

The moveable slider portion has a scale for a forward and rearward belly baggage compartment giving the weight in pounds in the upper left corner. A passenger scale is in the lower left hand corner giving the passengers by number, not by weight.

On the right-hand side are fuel tanks with the fuel weight for each tank given (Fig. 4-6E).

To compute a problem with this computer, the basic reference number and empty weight must be known. For this purpose, the weight of the aircraft will be 150,000 pounds and the index number will be 680. The zero fuel weight must be less than 230,000 pounds and the takeoff weight may not exceed 334,000 pounds.

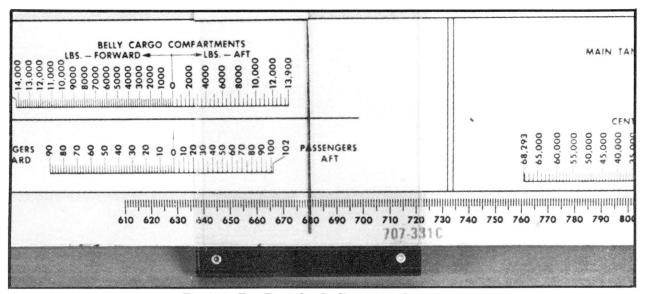

Fig. 4-6   F — Basic load adjuster setting.

Step 1:   Place the hair line of the cursor on 680 of the Basic Reference Number Scale (Fig. 4-6F).

Step 2:   Move the slider so the "0" mark of the passenger scale lines up with the cursor hairline. The hairline should still be lined

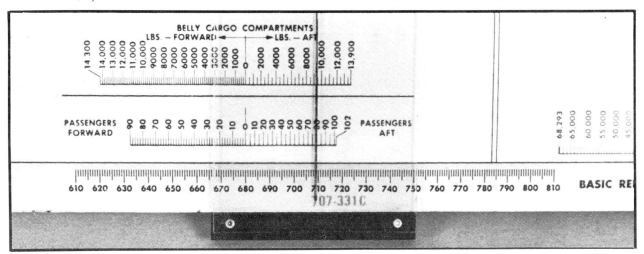

Fig. 4-6   G — Passenger loading.

up with the 680 index. Next move the cursor hairline to 80 passengers aft. The hairline should pass through the 710

mark on the basic reference number scale (Fig. 4-6G).

59

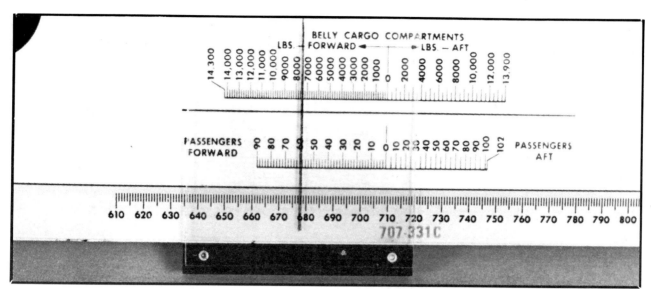

Fig. 4-6   H — Forward passenger loading.

Step 3: Move the slider to "0" on the passenger scale making sure the cursor remains on the 710 mark. Next, move the cursor to 60 passengers forward. The reference index should now read 678 (Fig. 4-6H).

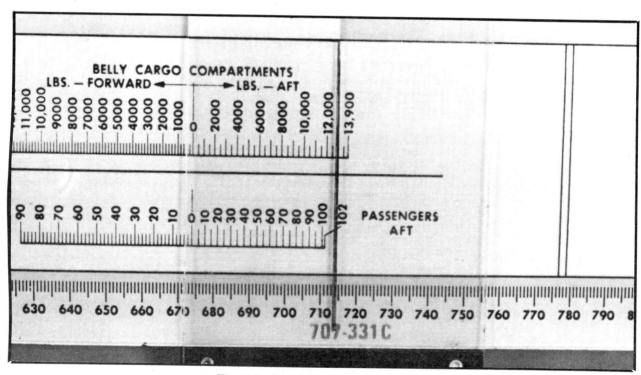

Fig. 4-6   I — Aft baggage loading.

Step 4: Next, using the belly cargo compartment scale, move the slider to the "0" mark on the cursor hairline. The hairline is still on the 678 mark of the basic reference number scale. Next, move the cursor so that the hairline passes through the 12,000 pound mark of the aft baggage. At this time, the hairline also passes through the 717 mark on the basic reference number scale (Fig. 4-6I).

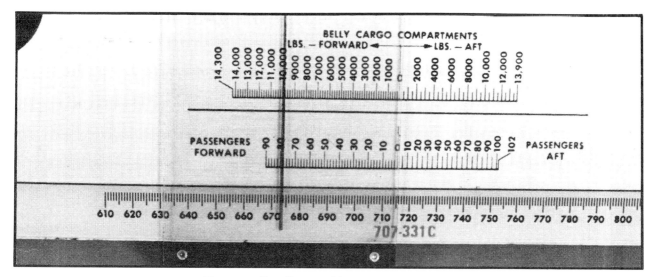

Fig. 4-6   J — Forward baggage loading.

Step 5: Move the slider to the "0" mark of the baggage compartment again. Next, move the cursor to the 10,000 pound mark of the forward baggage compartment. The hairline of the cursor should be on the 675 of the basic reference number index (Fig. 4-6J).

Step 6: Add the weights of the useful load to see if the maximum "0" fuel weight has been exceeded. The weight of passengers during the summer is based on 160 pounds.

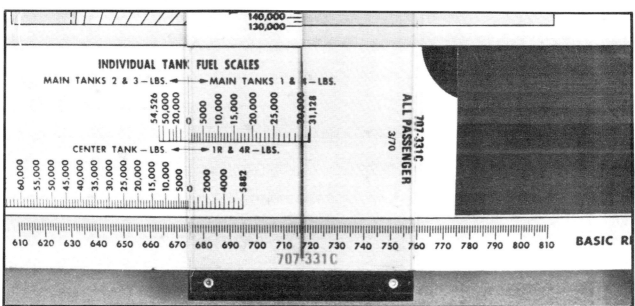

Fig. 4-6   K — Fuel in tanks 1 and 4.

Basic Aircraft Weight	150,000	pounds
80 Passengers Aft	12,800	pounds
60 Passengers Forward	9,600	pounds
Cargo Aft	12,000	pounds
Cargo Forward	10,000	pounds
Zero Fuel Weight	194,400	pounds

Step 7 Move the slider to the "0" mark of the main tanks 2 & 3 and 1 & 4. Make sure that the hairline is still at the basic reference number of 675. Move the cursor to 30,000 pounds in tanks 1 and 4. The hairline should now be on 717 (Fig. 4-6K).

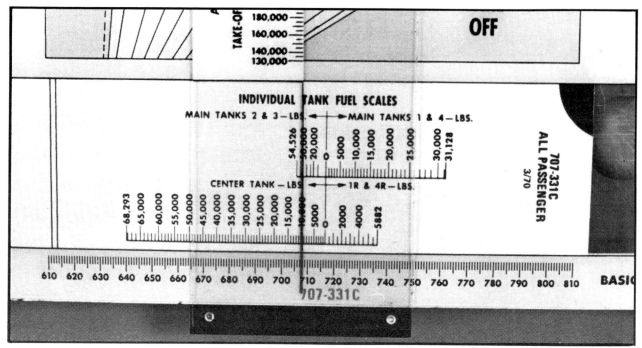

Fig. 4-6   L — Fuel in tanks 2 and 3.

Step 8: Move the slider back to the "0" mark of tanks 1 & 4 and 2 & 3. Now move the cursor to 50,000 pounds in tanks 2 and 3. The hairline should now pass through the 708 mark of the basic reference index (Fig. 4-6L).

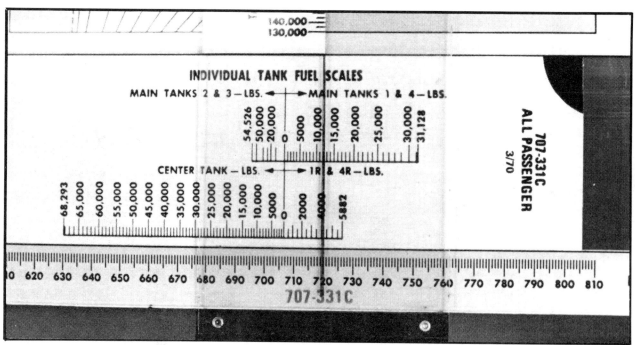

Fig. 4-6   M — Fuel in tanks center and 1 and 4 reserve.

Step 9: Move the slider so the "0" is on the hairline between the center tank and 1 and 4 reserve. Place the cursor on the 4,000 mark of 1 and 4 reserve. The hairline should pass through the 720 basic reference number (Fig. 4-6M).

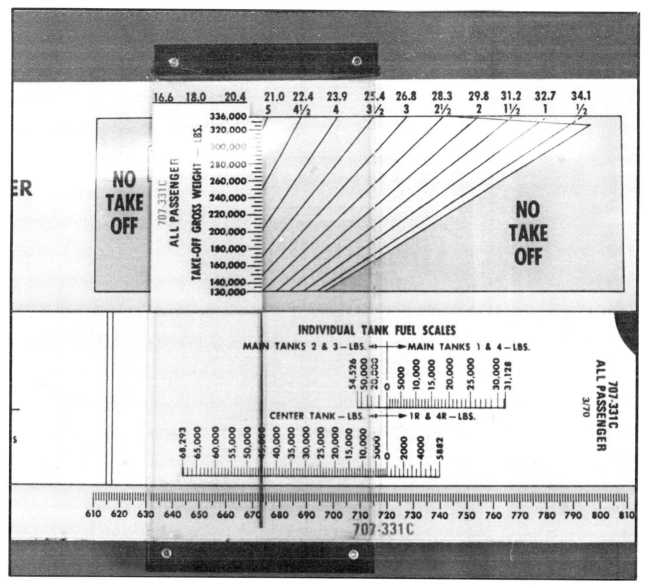

Fig. 4-6 N — Fuel in center tank.

Step 10: Move the slider to the "0" on the hairline between the center tank and 1 & 4 reserve. Place the cursor on 45,000 pounds. The hairline should be located at 674 (Fig. 4-6N).

Step 11: Add the Zero fuel weight to the weight of the fuel.

Zero Fuel Weight	194,400	pounds
Tanks 1 & 4	30,000	pounds
Tanks 2 & 3	50,000	pounds
Tanks 1 & 4 Reserve	4,000	pounds
Center Tank	45,000	pounds
Total Aircraft Weight	323,400	pounds

Note: The aircraft is below takeoff weight.

Step 12: With the hairline of the cursor on the basic index number of 674, look at the gross takeoff weight on the cursor scale. The MAC is about 21% and the nose up trim is around 5. The aircraft is within all limits.

E. Gear Retraction Moment

For most retractable geared airplanes, the center of gravity is calculated with the gear down. In some instances, the moment of the retraction can have an adverse effect on the center of gravity.

63

This, of course, is normally the nose gear that has the adverse effect because it will fold into the nose of the aircraft while most main gears travel in a lateral direction maintaining the same arm as the gear retracts (Fig. 4-7).

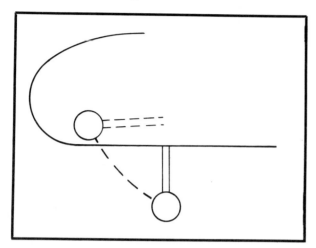

Fig. 4-7 Gear retraction change.

For this reason, the weight and balance section of the flight manual often specifies this fact when using graphs and tables. Often the moment of the retracted gear is also given so that it may be calculated in the center of gravity. This is done most often on large aircraft where the gear would be long and heavy. The following would be such a problem:

Takeoff Weight is 193,000 pounds.
Gear Down CG is 20.7% MAC.
LEMAC is 786 inches.
TEMAC is 1,086 inches.
Nose gear retracts forward with a moment of 268,000.

Step 1: The MAC is given in percent and must be converted to inches. Subtract TEMAC from LEMAC.

TEMAC	1086
LEMAC	−786
MAC	300

The MAC is 300 inches long or 300 inches is 100% of the MAC. Multiply 3.00 X 20.7 to obtain the present center of gravity in inches. 20.7 X 3.00 = 62.100 inches.

Step 2: The Center of gravity change is the nose gear moment divided by the weight or

$$\frac{268000}{193000} = 1.4.$$

The center of gravity moved 1.4 inches forward.

Step 3: Since the old center of gravity was located at 62.1 inches the new center of gravity will be located at 62.1 − 1.4 or 60.7 inches.

Step 4: The new center of gravity must be converted back to a percent. To do so the new CG must be multiplied by 100 and divided by the MAC or

$$\frac{60.7 \text{ X } 100}{300} = 20.2\% \text{ MAC.}$$

The gear retraction is often shown in graph form for easy calculation as shown in Fig. 4-8. If the MAC had not been given in a percent steps 1 and 4 would not have been needed.

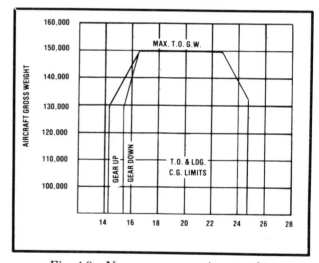

Fig. 4-8 Nose gear retraction graph.

F. Weight Shifting Formula

On occasion, after the loaded center of gravity is determined, the aircraft may be within the weight limitation but not within the center of gravity range. When such a situation does exist the load must be shifted from one location to another. This can be accomplished by a trial and error method but this is quite time consuming, especially if a large load of cargo is involved. For this reason a formula may be used to determine weight shift. The formula is as follows:

$$X = \frac{W \times D}{d}$$

W = Aircraft Maximum Weight

D = Distance the CG is to be shifted

d = Distance between stations in inches where weight is to be shifted.

This formula can be used anytime that any of the three factors are known. For example:

$$D = \frac{Xd}{W}$$

For this reason the same formula may be used for several other purposes such as determining the location of new equipment items, fuel consumption or moveable ballast locations.

PROBLEM 1

If a certain aircraft has a maximum takeoff weight of 70,000 pounds and the center of gravity was 2 inches out of forward limits and the cargo compartments are 100 inches from each other, how many pound of cargo must be moved?

Step 1: Using the formula assign the values to the letter in the formula:

W = 70,000 pounds

D = 2 inches

d = 100 inches

Step 2: Apply the numbers to the formula and solve the problem.

$$X = \frac{W \times D}{d} \text{ or } \frac{70,000 \times 2}{100}$$

$$= \frac{140,000}{100} = 1,400 \text{ pounds}$$

1,400 pounds must be shifted from the forward cargo compartment to the rear cargo compartment.

Step 3: The center of gravity should be rechecked using the Total Moment divided by the Total Weight Formula to determine that no error exists in the calculation.

PROBLEM 2

The helicopter used in Problem 1H had the center of gravity located at +3.75 and had an empty weight of 1,551 pounds and a maximum weight of 2,350 pounds. If it is loaded with 2 passengers weighing 170 pounds each, 1 pilot of 180 pounds, full fuel and oil, the aircraft will be out of forward CG as follows:

Negative Moments	Weight	Arm	Moment
Pilot	180	−30	−5400
2 Passengers	340	−30	−10200
Total	520		−15600

Positive Moments	Weight	Arm	Moment
Empty Aircraft	1551.0	+3.75	+5816.25
Fuel (29 gal.)	174.0	+24	+4176.00
Oil (3 gal.)	22.5	+5.0	+112.50
	1747.5		+10104.75
Pilot & Passengers	+520.0		−15600.00
	2267.5		−5495.25

The loaded CG is located at a −2.42. This is beyond the limits of the forward center of gravity. As fuel is consumed the aircraft will become more noseheavy. When the aircraft was weighed, the battery was located in the forward position. Possibly by moving the battery the aircraft may fall within the center of gravity limits. Since the fuel will change the center of gravity, the estimated fuel consumption will be 100 pounds.

Total Weight		Total Moment	
	2267.5		−5495.25
Fuel Consumed	−100.0		−2400.00
	2167.5		−7895.25

$$\text{New CG} = \frac{TM}{TW} \text{ or } \frac{-7895.25}{2167.5} = 3.6 \text{ inches}$$

Using the Formula $X = \frac{W \times D}{d}$, the amount of weight to be shifted can be determined.

Step 1: Using the formula assign the values to the letters.

65

W = 2350

D = 3.6 − 2 or 1.6 inches (3.6 is the new CG. The forward limit is 21 inches)

d = 67 + 96 or 163 inches (the battery was located at −67. The aft location is +96.)

Step 2: Apply the numbers to the formula and solve the problem.

$$X = \frac{W \times D}{d} \quad \text{or}$$

$$X = \frac{2350 \times 1.6}{163}$$

$$= 23.06 \text{ pounds}$$

Step 3: The battery weighs 30 pounds so the aircraft could be flown safely by placing the battery in the rear location. In an investigation of the center of gravity, it would be determined that with one pilot and one passenger the battery would be located in the front. With one pilot and two passengers the battery would be in the rear. The aircraft must be placarded in this manner.

G. Cargo Loading

In many ways the loading of cargo is as unique as are the items that are carried in the aircraft. Today many of our large aircraft are built especially for freight with containers, specially designed floors and wide doors. Still another group of aircraft are built for conversion from passenger to cargo by removeable seats and interiors. A few aircraft are built for combination loads of passengers.

When cargo is carried, several special considerations must be observed to insure the safety of the flight and the cargo carried. Some of these precautions are as follows:

1. Floor loading
2. Proper security
3. Center of gravity

All aircraft have floor limitations determining the load per square inch that may be applied to a section of the floor. Usually, for weight and balance purposes, the aircraft is divided into sections with maximum weights assigned to each section. This may also be for structural reasons as well (Fig. 4-9).

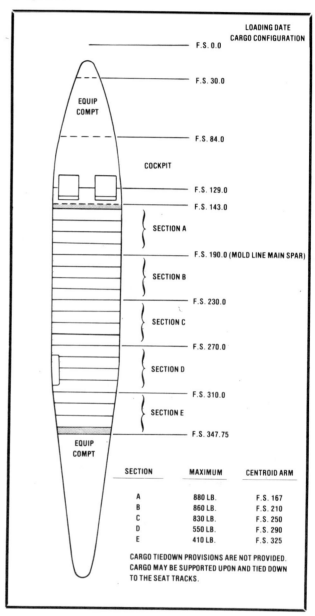

SECTION	MAXIMUM	CENTROID ARM
A	880 LB.	F.S. 167
B	860 LB.	F.S. 210
C	830 LB.	F.S. 250
D	550 LB.	F.S. 290
E	410 LB.	F.S. 325

CARGO TIEDOWN PROVISIONS ARE NOT PROVIDED. CARGO MAY BE SUPPORTED UPON AND TIED DOWN TO THE SEAT TRACKS.

Fig. 4-9 Cargo floor loading chart.

When extremely heavy objects or objects with heavy load concentrations are placed on the floor, shoring or planking may be necessary to evenly distribute the load. This may even necessitate leaving one adjacent area without cargo (Fig. 4-10).

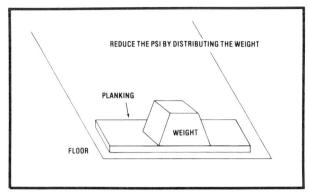

Fig. 4-10 Planking used to distribute cargo weight.

Security of all cargo is most important. Shifts in cargo may adversely affect the center of gravity or could possibly exit the aircraft through the fuselage. A number of different methods have been provided for securing cargo by the manufacturers, such as rings in the floor, a structure for straps and nets, locking devices to seat rails, and container locks to cargo rails (Fig. 4-11).

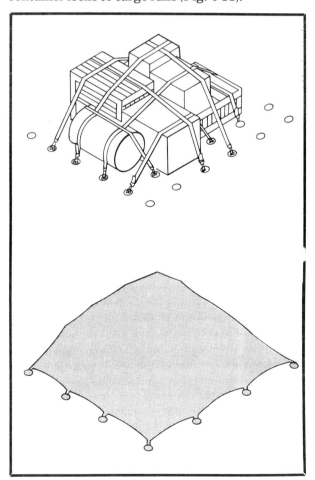

Fig. 4-11 Cargo tiedown methods.

In all cases when cargo is hauled, the weight and balance is most important. The weight should not be estimated. Certain items can present unique problems with the center of gravity. One such item is a liquid. If at all possible, all liquid containers should be carried with the cap up so that spillage does not occur. One 55 gallon drum leaking during a flight could change the center of gravity considerably in a light aircraft. When containers such as drums must be placed on their side, special care should be taken to see that they are full or that a smaller container is used. Remember that the movement of a liquid in takeoff can change the center of gravity.

At no time should loads be carried outside the aircraft unless such provisions have been made or the manufacturer consulted as to the feasibility. The aerodynamic characteristics may be affected even though the aircraft is within weight and balance limitations.

When a combination of cargo and passengers are carried in the same area, it is advisable to carry the cargo in front of the passengers.

In computing the cargo loads, the same systems mentioned previously may be used such as graphs, charts, sample loadings and computers. For this reason, no problems will be worked in this area.

1. Helicopter loading

For the most part, the weight and balance loading of the helicopter is the same as that of the fixed wing aircraft. However, it is more critical because of the very short center of gravity range present in most helicopters. The ideal place for the center of gravity would be directly under the main rotor mast. Since this is not possible, the range seldom travels more than a few inches in either direction from this location. On some helicopters, it is less than a total of three inches.

As a final check of the center of gravity, most helicopter pilots will hover the aircraft and check cyclic control movement in the fore and aft direction to determine if it is nose or tail heavy. To counteract such conditions in some helicopters, batteries have been placed on moveable racks and moveable ballast has been used. On more than one occasion the baggage compartment has been filled with items of weight by the pilot to compensate for weight changes.

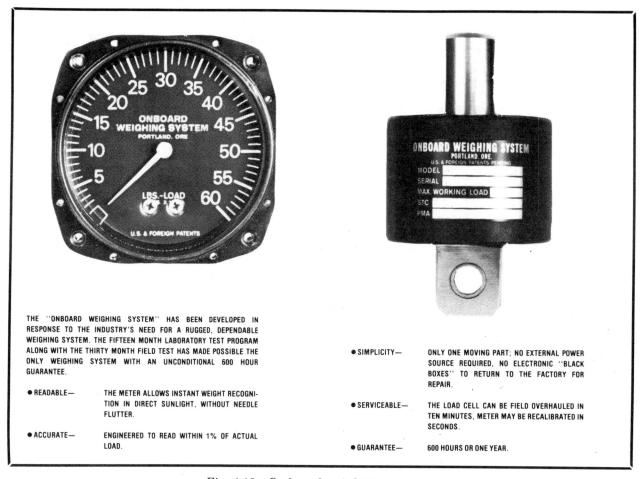

THE "ONBOARD WEIGHING SYSTEM" HAS BEEN DEVELOPED IN RESPONSE TO THE INDUSTRY'S NEED FOR A RUGGED, DEPENDABLE WEIGHING SYSTEM. THE FIFTEEN MONTH LABORATORY TEST PROGRAM ALONG WITH THE THIRTY MONTH FIELD TEST HAS MADE POSSIBLE THE ONLY WEIGHING SYSTEM WITH AN UNCONDITIONAL 600 HOUR GUARANTEE.

● READABLE— THE METER ALLOWS INSTANT WEIGHT RECOGNITION IN DIRECT SUNLIGHT, WITHOUT NEEDLE FLUTTER.

● ACCURATE— ENGINEERED TO READ WITHIN 1% OF ACTUAL LOAD.

● SIMPLICITY— ONLY ONE MOVING PART; NO EXTERNAL POWER SOURCE REQUIRED, NO ELECTRONIC "BLACK BOXES" TO RETURN TO THE FACTORY FOR REPAIR.

● SERVICEABLE— THE LOAD CELL CAN BE FIELD OVERHAULED IN TEN MINUTES, METER MAY BE RECALIBRATED IN SECONDS.

● GUARANTEE— 600 HOURS OR ONE YEAR.

Fig. 4-13   On board weighing system.

The weight lifted by the helicopter is also very critical, especially when loads are carried from one density altitude to another, such as a mountainous terrain. As the density altitude increases, the ability of the helicopter to hover becomes much less. For this reason a helicopter pilot must be careful about landing at a higher altitude than the one from which he took off (Fig. 4-12).

Some helicopters are involved in external loads. This type of operation is covered in FAR 133 which requires standards for external loading. Usually the weight limitations for external loads are the same as for internal loads. For this reason helicopters are used exclusively for external loads and kept at a minimum empty weight.

The hook for external loads is placed as close to the ideal center of gravity as practical. These cargo hooks have two releases so that the possibility of not being able to release the load is kept to a minimum. If the load becomes dangerous or other difficulties are experienced, the load may be jettisoned by either release mechanism, one of which is mechanical in case of electrical failure.

Few problems should be encountered with the CG because of the proximity to the ideal center of gravity if the load remains stationary.

However, great care must be exercised with external loads because of oscillating tendencies of loads in forward flight and the possibility of a pendulum effect on a load.

When the weights are known, very few problems will be encountered with the weight if the maximum weight is not exceeded and the density altitude charts are adhered to. When unknown weights are lifted, difficulties may be encountered. One device which is used to avoid such problems is an onboard weighing system. A load cell is placed between the hook and the load giving an actual weight on a gauge in the cockpit (Fig. 4-13).

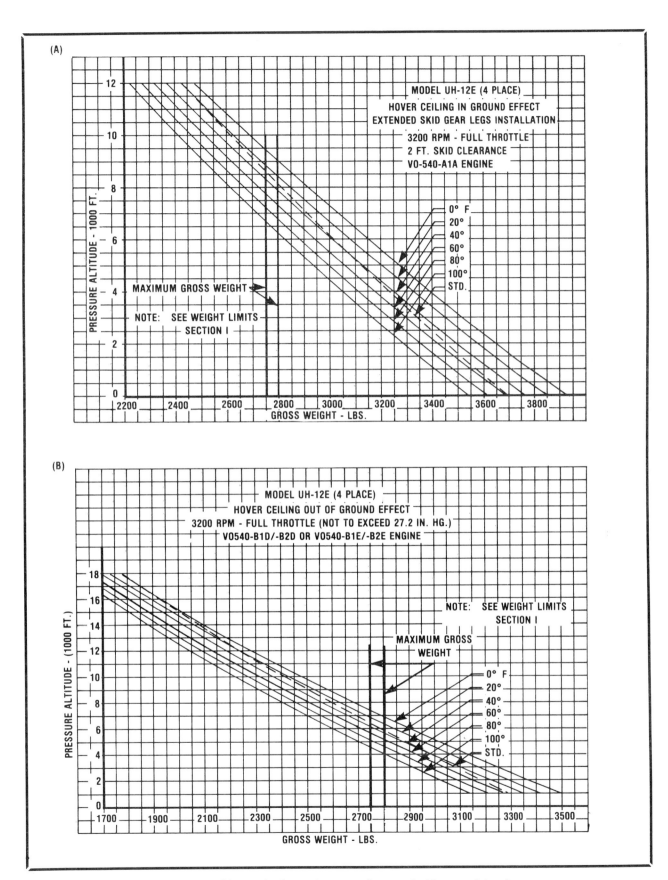

Fig. 4-12   Charts for hovering out of ground effect at altitude.

69

H. Air Taxi

All aircraft operated under FAR 135, referred to as Air Taxi, have a number of privileges and requirements that do not apply to privately owned aircraft in general use.

One of these requirements is that all multi-engine aircraft used for air taxi must be physically reweighed *every* 36 calendar months. This requirement, however, does not apply to new aircraft less than 36 months old even though the manufacturer may not have weighed that particular aircraft but merely calculated the weight at the time of production. If the aircraft was purchased from another operator who had the aircraft reweighed it is not necessary that the new operator reweigh the aircraft before using it in service, *but it must be reweighed at the 36 month period.* For example, if the aircraft was purchased 24 calendar months after the last reweighing, it will require reweighing 12 calendar months after purchase.

When the aircraft has had an accumulated weight change of one-half of one percent MAC, a new weight and balance should be established regardless of the time period.

For a group of aircraft of the same model and configuration a fleet empty weight may be used to compute the center of gravity.

The fleet weight is the average weight of the aircraft. For three of the same aircraft, all must be weighed. For four to nine aircraft, three aircraft must be weighed, and 50% of the number above three. For a fleet of nine or over, six aircraft must be weighed plus ten percent of the number over nine.

It is suggested that when choosing the aircraft to be weighed for fleet weights, that the aircraft with the highest time since the last weighing be used. If the weights of any of these aircraft vary more than + or − one-half of one percent in empty weight or center of gravity, the individual aircraft should be operated at its actual weight and center of gravity. For those aircraft remaining that are not physically weighed, the center of gravity and weight are simply computed from the average fleet weight.

Usually, the weighing procedure is the same as other aircraft except the scales are to be certified within one year.

A *loading schedule* must be kept with the aircraft. This loading system may be of any form as long as it is simple and orderly. For example, it may be the manufacturers system, a chart system, load adjuster or sample loadings. Regardless of the system it must take into account the fuel, oil, baggage, gear retraction and passenger movement.

The fuel portion must include not only the initial fueling but the fuel consumed by the aircraft in flight.

If baggage restrictions are required, these restrictions should be marked in the appropriate area as well as being included in the loading areas.

In calculating for passenger movement, the most adverse condition should be used. This might be done by movement of the flight attendant, or occupancy of the lavatory.

In the computation of the passenger load, certain seats may be blocked off to insure a proper center of gravity location. The operator may also elect to use actual weights or average weight as long as the passengers are obviously not above or below the average weight. For example, a football team. When using average weights, they will be as follows:

Adults: 160 pounds (Summer) from May 1 to October 31
165 pounds (Winter) from November 1 to April 30

Children: 2 - 12 years -- 80 pounds
Over 12 years -- Adult
Under 2 -- Babes in Arms

Crew: Female Attendents: 130 pounds
Male Attendents: 150 pounds
Other Crew: 170 pounds

Baggage
Weights: Checked in: 23.5 pounds
Carried on: 5 pounds

NOTE: Charter or special purpose flights are not included in the average baggage weights.

A load manifest is required for all multi-engine aircraft operated under FAR 135. The load manifest must include the following information:

1. Number of passengers
2. Total weight of the loaded aircraft
3. Maximum allowable takeoff weight
4. Center of gravity of the loaded aircraft
5. Center of gravity limits of the aircraft

A copy of the load manifest must be carried on the flight and kept by the operator for a period of 30 days (Fig. 4.14).

I. Addition and Removal of Equipment

From time to time throughout the life of an aircraft, new equipment is installed and old equipment is removed.

This new equipment may be nothing larger than a clock or it may be an entirely different engine installation. In any case, this installation or removal will result in a change in the center of gravity and some additional paperwork.

One type of change of equipment may be a kit which will either be an additional item or items approved by the manufacturer in a kit form or a kit manufactured by someone other than the manufacturer that has received a *Supplemental Type Certificate* (STC) for installation in a certain type of aircraft. Either type of installation is usually quite simple from a weight and balance standpoint because the weight and the arms of the items are given in the installation instructions unless the equipment installation is large or complex. In these cases the kit manufacturer will suggest using the actual weight which will result in reweighing the aircraft, calculating the new center of gravity and correcting the equipment list.

Some installations of equipment will not come in a kit form. This could be the installation of a radio, extra instrumentation, etc. These types of installations will normally not be approved by the manufacturer but have been built to a *Technical Standard Order* (TSO) insuring its airworthiness. Generally the weight of TSO'd items are given on the item itself. Since no location for the item has been predetermined, the arm will not be given. In such a situation, the distance of the arm will actually have to be measured.

This could be a rather difficult problem when the item is installed inside the aircraft. To simplify the matter the arm is usually measured from a known item or station number.

The other types of installations are usually of a *one of a kind* type such as special equipment, a different door system, etc., which are not manufacturer approved items or covered by a Supplemental Type Certificate or TSO'd. Since they would result in a weight change and a change in the center of gravity, these types of changes often require *Field Approval* which can be given by the FAA. They will also normally result in reweighing the aircraft if the modification is very complex.

In any case, any item that is removed or installed requires a revision of the equipment list and the new center of gravity must be computed.

It is very important that the old center of gravity be marked *"Superseded (and the date)"*, so that no more than one center of gravity exists for an aircraft at any one time.

All changes of the center of gravity must be recorded in the aircraft log and are generally placed on a paper attached to the equipment list with other pertinent weight and balance data. It is required that all aircraft have current weight and balance data.

For changes of equipment, the original weight and balance data from the last actual weighing will usually be used in conjunction with the total moment divided by the total weight formula.

For an example, the 310 L used in the review problem at the end of the chapter will be utilized. The total empty weight was 3,419 pounds with a moment of 120756.3 locating the center of gravity at 35.3 inches.

The first change to the weight and balance will be the removal of the 3 bladed propellers which are optional equipment and replace them with the standard two blade propellers.

Step 1: Remove the 3 bladed propellers that weigh 166 pounds on a −28 arm and the de-icer system that weighs 13 pounds with an arm of −14.0.

71

PASSENGER — CARGO LIST	WEIGHT	SEAT COMPT	INDEX
	180	1	126.0
	170	2	119.0
	140	3	126.0
PASSENGER NAMES	150	4	135.0
	200	5	220.0
ENTERED HERE	160	6	176.0
	210	7	273.0
	130	8	169.0
	120	9	180.0
	140	10	210.0
	130	11	221.0
PASSENGER — CARGO — MAIN CABIN TOTAL			

	FUEL					
WEIGHT	TANK	INDEX	BAGGAGE COMPT	100	NOSE	20.0
300	R—MAIN	270.0	BAGGAGE COMPT	200	REAR	380.0
300	L—MAIN	270.0	PILOT—COPILOT	330		165.0
120	R—AUX	144.0	OIL	85		7.7
120	L—AUX	144.0	FUEL	840	TOTAL	828.0
	NOSE		EMPTY WEIGHT	6,150		5,842.5
840	TOTAL	828.0	TAKEOFF WEIGHT	9,435		9,198.2
			TAKEOFF LIMITS	10,000		8467.9 9694.5

AIRCRAFT   N 123 QC

FLIGHT   45

ROUTE   OKC — MKC

DATE _____

I HEREBY CERTIFY THAT THE ABOVE TAKEOFF WEIGHT AND INDEX IS WITHIN LIMITS AS SPECIFIED IN THE FLIGHT OPERATIONS MANUAL

_____

PILOT IN COMMAND

Fig. 4-14 *Sample load manifest.*

	Weight	Arm	Moment
3 Bladed Propellers	166 lbs.	−28	−4648
3 Bladed De-icer System	13 lbs.	−14	−182
	179 lbs.		−4830

Step 2: Install 2 bladed propellers that weigh 133.7 pounds at −28 inches and de-icer system that weighs 12.4 pounds at −13.4 inches.

	Weight	Arm	Moment
2 Bladed Propellers	133.7 lbs.	−28.0	−3744
2 Bladed De-icer System	12.4 lbs.	−13.4	−166
	146.1		−3910

Step 3: Subtract the weight and moment of the 2 bladed propellers from the 3 bladed propellers and moment.

	Weight	Moment
	179.0 lbs.	−4830
	146.1	−3910
	32.9	+920

Step 4: Remove 32.9 pounds from the Empty Weight and add the difference of the moment.

	Weight	Moment
Empty Aircraft	3419.0	120756.3
	−32.9	+920.0
	3386.1	121676.3

Step 5: Divide the new total moment by the new total weight in order to obtain the new CG.

$$\frac{121676.3}{3386} = 35.9$$

The new CG is 35.9. This will be recorded in the log and the equipment list will be revised with the two bladed propellers added and the three bladed propellers removed from the list. Be sure the old weight and balance is superceeded.

Next, the aircraft will have a Collins PN - 101 Pictoral Navigation System added. Since this is approved equipment by the manufacturer their recommendations will be followed. The system weighs 22.7 pounds at an arm of 117.8 inches.

Step 1: Multiply the weight times the arm to determine the moment.

$$22.7 \times 117.8 = 2674$$

Step 2: Add the weight and moment to the empty weight and moment.

Empty Aircraft	3386.1	Empty Moment	121676.3
	22.7		2674.0
	3408.8		124350.3

Step 3: Divide the total moment by the total weight to determine the new CG.

$$\frac{124350.3}{3408.8} = 36.47$$

The new center of gravity is 36.47. Enter this in the logbook, supercede the old weight and balance and add the item to the equipment list.

1. Ballast procedure

Some aircraft, especially after modification or removal of equipment, will no longer have a favorable location for the center of gravity. This situation may be a temporary or a permanent problem depending upon how it was caused.

If it were caused by the removal of a piece of equipment for repair that would be replaced shortly, and which did not otherwise affect the airworthiness of the aircraft, or if a certain loading condition was needed for a specific flight, temporary ballast could be used. This could be in several different forms, but probably the best is lead shot in bags or bars. However, any object of weight could be used. All ballast of this type should be placarded as: *Ballast (?) lbs. − Removal Requires Weight and Balance Check.* This will help insure that it is not inadvertently removed from the aircraft by anyone not knowing its purpose. Such ballast is often placed in the baggage area. In all cases, this temporary ballast should be secured so that no dangerous shift of weight may occur.

If the center of gravity has been relocated to an adverse location permanently, it would be best if some equipment could be relocated such as a battery, inverter or other items of weight. By relocation of equipment, no change will be made in the useful load, nor will any additional weight be required.

When it is not feasible to relocate equipment in order to obtain a favorable center of gravity in the loaded condition, it will become necessary to permanently ballast the aircraft. This means that additional weight will be added to the aircraft at predetermined locations. This weight will become a part of the empty weight of the aircraft and will remain a permanent part of the aircraft as long as it is flown in its present condition. Since this is not a favorable condition, it should be avoided if at all possible because it will reduce the useful load of the aircraft.

Lead is usually used for ballast because of its density. However, other metals such as iron and brass have been used. Whatever material is used, it must be marked *"Permanent Ballast. Do not remove"*.

The most ideal position for ballast would be on the longest arm possible or in the nose or the tail depending upon the moment that is being counteracted. Unfortunately, it is not always possible to locate the ballast in these extreme locations because of structural weight considerations (Fig. 4-15). For this reason, it is often necessary to compromise by using an area of the aircraft with a shorter arm location, but that has the structural capability of supporting the weight of the ballast. Such a place is often the baggage area of the aircraft or unused radio racks because very few aircraft already have provisions made for ballast. Of course, anytime that the arm is shortened, the weight of the ballast must be increased which decreases the useful load of the aircraft.

On some aircraft, especially helicopters, provisions are provided. It is best to find a location that will have the unquestionable structural integrity to support the weight which is why the baggage areas were suggested. Often ballast is needed because of intensive modifications to the aircraft. This sometimes leaves areas available

that have previously supported equipment items of a known weight that were removed from the aircraft in order to support the ballast. If no such locations are available it may be necessary to construct a suitable support for the ballast. It must be remembered that the support must not only be capable of supporting the static weight but the additional "G" forces that may be imposed during flight and landing. The following is a chart of the static load required to support a 7 pound weight in the utility category.

Load Factors	Static Test Load X 7
Sideward 1.5 G	10.5 pounds
Upward 3.0 G	21.0 pounds
Downward 6.6 G	46.2 pounds
Forward 9.0 G	63.0 pounds

Further information about construction of supports is furnished in FAA AC 43.13.2 on Alterations.

The ballast to be installed should be kept to a minimum so that no more weight is carried on the aircraft than necessary in order to keep the center of gravity within limits. It would be possible to use a trial and error method of adding weight and determining the center of gravity but this would be a very slow process. There are several formulas available for determining the amount of ballast required. The formula that will be used in this text is as follows:

$$\text{Ballast} = \frac{\text{Derived Weight (Required CG} - \text{Derived CG)}}{\text{Ballast Arm} - \text{Required CG}}$$

This formula will work for all ballast problems that may be incurred such as a loaded center of gravity location or an empty center of gravity location.

Probably no type of aircraft utilizes ballast as frequently as helicopters. This is mainly due to the short center of gravity range and some of the specialized equipment that they may be required to carry for work that is performed. For this reason many helicopters have provisions built into the airframe to accommodate ballast. The Bell Jet Ranger used in Problem 2H has this type of provision. It might also be noted that the empty center of gravity was not within the graph limits

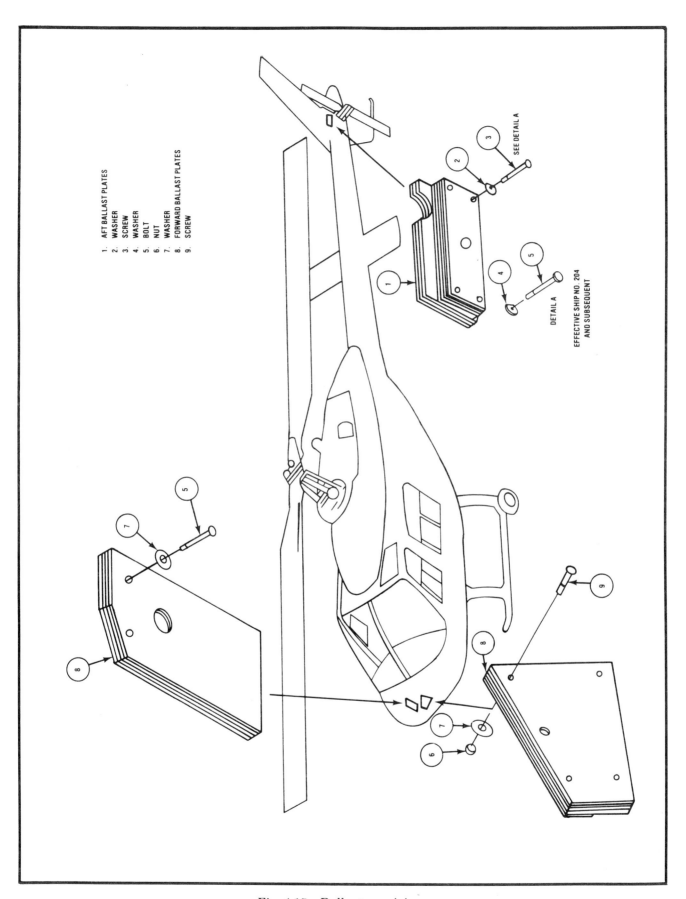

1. AFT BALLAST PLATES
2. WASHER
3. SCREW
4. WASHER
5. BOLT
6. NUT
7. WASHER
8. FORWARD BALLAST PLATES
9. SCREW

SEE DETAIL A

DETAIL A

EFFECTIVE SHIP NO. 204
AND SUBSEQUENT

Fig. 4-15   Ballast provisions.

for operating without limitations under normal loading conditions. In this situation the aircraft could be placarded or ballast may be added to obtain a more favorable center of gravity. It would be more advantageous in most helicopter operations to ballast than to placard. Using the information obtained in Problem 2H and the ballast formula the correct amount will be calculated.

The Empty CG is 118.83 inches
The Empty Weight is 1557.2 pounds
The Ballast Locations are (+13) and (+341.63)
The Empty Moment is 185042.1

Step 1: Using the graph in Fig. 3-19, the center of gravity should be located at approximately 117.84 inches from the datum.

Step 2: Since the aircraft exceeds the rearward limit the ballast must be placed at the (+13) location to move the center of gravity forward.

Step 3: With all the information known for the formula, it is placed in the appropriate sections of the formula:

$$Ballast = $$

$$\frac{Derived\ Weight\ (Required\ CG - Derived\ CG)}{Ballast\ Arm - Required\ CG}$$

$$Ballast = $$

$$\frac{1557.20\ (117.84 - 118.83)}{13 - 117.84}$$

Step 4: Subtract the required CG from the derived CG and the Ballast Arm from the required CG. They should be .99 and 104.84 respectively or

$$Ballast = \frac{1557.20 \times .99}{104.84}$$

Step 5: Multiply the derived weight by .99 and divide by 104.85 or

$$\frac{1541.62}{104.84} = 14.7 \text{ pounds of ballast.}$$

NOTE: For our purposes 15 pounds of ballast will be used.

Step 6: Always check the ballast installed by recalculating the center of gravity using the new weight and moment. $1557.2 + 15 = 1572.2$.

The moment of the ballast is weight X arm or $15 \times 13 = 195$.

The old moment was 185042.1.

The new moment is $185042.1 + 195$ or 185237.1.

Using the formula $\frac{TM}{TW}$ or $\frac{185237.1}{1572.2}$
= 117.82 (New CG.)

The center of gravity is now within the limits of the Empty CG Range.

QUESTIONS:

1. Who is responsible for the weight and balance of the aircraft.

   A. Owner.

   B. Pilot.

   C. Maintenance personnel that calculated the weight and balance.

2. Minimum fuel is used in the calculation of:

   A. Empty center of gravity.

   B. The most forward CG.

   C. The most forward or rearward CG depending upon the tank location.

3. All minimum weights are used aft of the rearward limit when the most forward CG is calculated.

   A. True.

   B. False.

4. A loading chart may be in the form of:

   A. A placard.

   B. A graph.

   C. Either of the above.

5. When graphs are used, the moment is normally reduced.

   A. True.

   B. False.

6. Fuel consumption will not affect the center of gravity during flight.

   A. True.

   B. False.

7. Landing weight may be the same as the takeoff weight on all large aircraft.

   A. True.

   B. False.

8. When using the load adjuster, the same index number is used for the same type of aircraft.

   A. True.

   B. False.

9. When retracting the gear on large nosewheel aircraft, the center of gravity will normally move:

   A. Aft.

   B. Forward.

   C. Will not move.

10. When shifting weight in order to obtain a favorable center of gravity:

   A. The aircraft maximum weight must be known.

   B. The distance that the CG is to be moved must be known.

   C. The distance that the weight is to be moved must be known.

   D. All of the above must be known.

11. When loading heavy items, it is sometimes necessary to use planking. This is for the purpose of:

   A. Keeping the load from shifting.

   B. Because of floor loading.

   C. It is never necessary.

12. The helicopter pilot's final check of the CG is:

   A. Bringing the helicopter to a hover.

   B. Checking cyclic travel.

   C. Both of the above.

13. External loads may be carried by all helicopters without restrictions.

   A. True.

   B. False.

14. Multi-engine aircraft used for Air Taxi must be reweighed every:

   A. 36 calendar months.

   B. 24 calendar months.

   C. The statement is false. All Air Taxi must be reweighed.

15. Fleet weight is used by the manufacturer in order to obtain the average weight.

    A. True.

    B. False.

16. The average weights used by FAR 135 operators are the same as the standard weights used in computing the most forward and rearward CG.

    A. True.

    B. False.

17. The addition of equipment requires an addition to the equipment list only.

    A. True.

    B. False.

18. Ballast may be placed anywhere in the aircraft.

    A. True.

    B. False.

19. When installing the ballast, the length of the arm is most important.

    A. True.

    B. False.

20. Old weight and balance calculations:

    A. Should be disposed of.

    B. Marked "Superceded".

GLOSSARY

This glossary of terms is provided to serve as a ready reference for the words with which you may not be familiar. These definitions may differ from those of standard dictionaries, but are in keeping with shop usage.

aircraft listings Information sheets published by the FAA containing pertinent information on a particular model aircraft — same as Specification Sheets except less than 50 aircraft are in existance.

aircraft specifications Information sheets published by the FAA containing pertinent data on a particular model of aircraft to include approved equipment.

aircraft specification data sheet Information sheets published by the FAA containing pertinent information on a particular model of aircraft. No information is contained on approved equipment.

arm The horizontal distance of any object from the Datum. It is expressed in inches and may be negative or positive depending on its relationship to the Datum.

ballast Weight used to obtain a favorable center of gravity location. It is usually lead and must be marked as such. It may be moveable or permanent.

basic empty weight See basic weight.

basic weight Weight of the aircraft to include unusable fuel and full oil, excluding all other useful load — sometimes referred to as basic empty weight.

category A group of the same type of aircraft according to usage. Ex: normal or utility category.

center of gravity The point at which the nose and tail moments are of equal magnitude. It is abbreviated CG and is sometimes shown symbolically as ⊕.

center of gravity envelope The forward and rearward limits of the aircraft as expressed in a graphic form.

center of gravity range The most forward and rearward CG limits in which the aircraft may be safely operated as designated by the manufacturer. This may be abbreviated CGR.

chord An imaginary line passing from the leading edge to the trailing edge of an airfoil.

data sheets See aircraft specification data sheets.

datum An imaginary vertical line from which all horizontal measurements are made or indicated. The datum may be located at any convenient location by the manufacturer.

empty center of gravity range The most forward and rearward limits of the empty aircraft as determined by the manufacturer. When these limits are met, no loading limits are necessary for standard loads. It may be abbreviated ECGR.

empty weight The weight of the aircraft with unusable fuel and no useful load.

equipment Any item that is secured in a fixed location to the aircraft to be utilized in the aircraft.

equipment list A comprehensive list of equipment installed on a particular aircraft. This includes the required and optional equipment.

fleet weight The average weight of several aircraft of the same model and with the same equipment. This weight may be used by 121 and 135 operators.

flight manual Information on flighting the aircraft as published by the manufacturer.

gross weight Maximum weight.

item A piece of equipment installed on an aircraft.

jackpoints Fixed locations on the aircraft to accommodate jacks for raising the aircraft. Sometimes referred to as "jack pads".

landing weight The weight of the aircraft at touchdown. This is often limited to less than the takeoff weight by the manufacturer for structural reasons.

lateral From side to side or from wing tip to wing tip.

LEMAC Abbreviation for leading edge of mean aerodynamic chord.

leveling lugs Points on the aircraft on which a level may be placed for leveling the aircraft.

leveling means The way in which the aircraft may be checked for level. This may be longitudinal or lateral and sometimes both.

leveling scale A scale built in the aircraft for checking the leveling of the aircraft in conjunction with a plumb bob.

load manifest An itemized list of weights and moments of a particular load taken on a specific flight. Used by 121 and 135 operators.

loading chart A chart of pre-calculated weights and moments.

longitudinal Lengthwise from the nose to the tail.

MAC Abbreviation for mean aerodynamic chord.

mean aerodynamic chord The average chord of an irregularly shaped wing. Often the CG is expressed in a percentage of MAC.

METO horsepower Maximum except takeoff horsepower. This is used in calculating minimum fuel.

minimum fuel An amount of fuel used when computing adverse loading for most forward and rearward conditions.

moment The product of the weight multiplied by the arm. The moment may be positive or negative depending on the sign of the weight and arm.

moment index Moment reduced by 100,000 or 10,000 for ease in balance calculation.

moveable ballast See ballast.

maximum takeoff weight The greatest designed weight at the time of lift off.

maximum weight Maximum weight allowed for a specific aircraft in a category. This is sometimes referred to as gross weight.

maximum zero fuel weight Maximum weight of the aircraft including useful load excluding the weight of the fuel. Some manufacturers designate this weight for structural reasons.

normal category An aircraft which is designed to perform normal maneuvers.

operating weight Term used by aircraft operators to include the empty weight of the aircraft and items always carried in the aircraft, such as crew, water, food, etc.

optional equipment Items installed on an aircraft that are not required for airworthiness.

payload Maximum design, zero fuel weight minus basic empty weight.

permanent ballast See ballast.

placard An information card contining information on the approved flight of the aircraft. Example "*Solo from the front seat only*".

ramp weight See taxi weight.

required equipment Items necessary on a type of aircraft to be airworthy.

station An imaginary vertical line denoting a horizontal distance from the datum expressed in inches.

takeoff weight Weight of the aircraft at lift off. Often used as maximum takeoff weight.

tare Any material used during the weighing process between the scale and the aircraft. Example: chocks.

taxi-weight Maximum weight allowed for ground maneuvering. Sometimes referred to as ramp weight.

TEMAC Abbreviation for trailing edge of the mean aerodynamic chord.

transport category Aircraft in excess of 12,500 pounds designed to perform normal maneuvers.

trapped fuel See undrainable fuel.

undrainable fuel Amount of fuel that remains in the system after draining. This is considered a part of the empty weight of the aircraft.

undrainable oil Oil remaining after draining the oil from an engine. This oil is considered a part of the empty weight of the aircraft.

unusable fuel Fuel that cannot be consumed by the engine. This fuel is considered a part of the empty weight.

unusable oil Oil that cannot be used by the engine. This unusable oil is not a part of the empty weight of the aircraft.

usable fuel Portion of the total fuel available for consumption by the aircraft in flight.

useful load Difference between the empty weight of the aircraft and the maximum weight of the aircraft. Sometimes referred to as the "payload".

utility category An aircraft designed to perform normal maneuvers and limited acrobatics.

zero fuel weight The weight of the aircraft to include all useful load except fuel.

FORMULAS

1. **Moment Formula:**

    Weight X Arm = Moment
    + Arm = if measured aft of Datum.
    − Arm = if measured forward of Datum.
    Positive X Positive = Positive
    Negative X Negative = Positive
    Positive X Negative = Negative
    Negative X Positive = Negative

2. **Center of Gravity:**

    $$CG = \frac{Total\ Moment}{Total\ Weight}$$

    W = Weight of Aircraft.
    D = Distance from Datum to the main wheel weighing point.
    L = Distance measured from the main wheel weighing point to the nose or tail wheel weighing point.
    F = Weight at the nosewheel weighing point.
    R = Weight at the tail wheel weighing point.

    (a) $CG = D - \dfrac{F\ X\ L}{W}$ = Nosewheel with Datum forward of the main wheels.

    (b) $CG = -D + \dfrac{F\ X\ L}{W}$ = Nosewheel with Datum aft of the main wheels.

    (c) $CG = D + \dfrac{R\ X\ L}{W}$ = Tail wheel with Datum forward of the main wheels.

    (d) $CG = -D + \dfrac{R\ X\ L}{W}$ = Tailwheel with datum aft of the main wheels.

3. **Lateral Center of Gravity:**

    $$CG = \frac{(AL\ X\ C) + (BR\ X\ D)}{W}$$

    W = Weight of the aircraft.
    AL = Butt Measurement Left.
    BR = Butt Measurement Right.
    C = Weight of main scale Left.
    D = Weight of main scale Right.

    Butt Line Left is Negative.
    Butt Line Right is Positive.

4. **Minimum Fuel:**

$$\text{Minimum fuel in gallons} = \frac{\text{METO}}{12}$$

$$\text{Minimum fuel in pounds} = \frac{\text{METO}}{2}$$

5. **Ballast:**

$$\text{Ballast} = \frac{\text{Derived Weight (Required CG} - \text{Derived CG)}}{\text{Ballast Arm - Required CG}}$$

6. **Weight Shifting Formula:**

$$X = \frac{\text{W X D}}{\text{d}}$$

X = Amount of Weight to be shifted.

W = Aircraft Maximum Weight.

D = Distance the CG is to be shifted.

d = Distance in inches where the weight is to be shifted.

7. **Gear Retration Formula:**

$$\text{CG Movement} = \frac{\text{Nose Gear Moment}}{\text{Aircraft Weight}}$$

STANDARD WEIGHTS

Gasoline: 6 pounds per gallon.

Turbine Fuel: 6.7 pounds per gallon.

Lubricating Oil: 7.5 pounds per gallon.

Crew and Passengers: 170 pounds per person.

Water: 8.35 pounds per gallon.

FAR 135 OPERATIONS

Adults: 160 pounds (summer) from May 1 to Oct. 31.
165 pounds (winter) from Nov. 1 to April 1.

Children: 2 - 12 years: 80 pounds.

Crew: Female Attendents: 130 pounds.
Male Attendents: 150 pounds.
Other Crew: 170 pounds.

Baggage: Check in: 23.5 pounds.
Carry on: 5 pounds.

MINIMUM FUEL TABLE		
Horsepower	**Gallons**	**Weight in lbs.**
65	5.416	32.50
70	5.833	35.00
75	6.25	37.50
80	6.66	40.00
85	7.08	42.50
90	7.50	45.00
95	7.91	47.50
100	8.33	50.00
125	10.41	62.50
150	12.50	75.00
175	14.58	87.50
200	16.66	100.00
225	18.74	112.50
250	20.83	125.00
275	22.91	137.50
300	25.00	150.00
325	27.08	162.50
350	29.17	175.00
375	31.25	187.50
400	33.32	200.00
425	35.41	212.50
450	37.50	225.00
475	39.58	237.50
500	41.66	250.00
525	43.74	262.50
550	45,82	275.00
575	47.90	287.50
600	49.98	300.00
625	52.06	312.50
650	54.14	325.00
675	56.22	337.50
700	58.30	350.00
725	60.38	362.50
750	62.46	375.00
775	64.54	387.50
800	66.62	400.00
825	68.70	412.50
850	70.78	425.00
875	72.86	437.50
900	74.94	450.00
925	77.02	462,50
950	79.10	475.00
975	81.18	487.50
1000	83.26	500.00

SPECIFIC GRAVITY CONVERSION TABLE			
Specific Gravity	**Pounds per U.S. Gallon**	**Specific Gravity**	**Pounds per U.S. Gallon**
.645	5.368	.685	5.702
.646	5.376	.686	5.710
.647	5.385	.687	5.718
.648	5.393	.688	5.727
.649	5.402	.689	5.735
.650	5.410	.690	5.743
.651	5.418	.691	5.752
.652	5.426	.692	5.760
.653	5.435	.693	5.768
.654	5.443	.694	5.777
.655	5.452	.695	5.785
.656	5.460	.696	5.793
.657	5.468	.697	5.802
.658	5.476	.698	5.810
.659	5.485	.699	5.818
.660	5.493	.700	5.827
.661	5.502	.701	5.835
.662	5.510	.702	5.843
.663	5.518	.703	5.852
.664	5.526	.704	5.860
.665	5.535	.705	5.868
.666	5.543	.706	5.877
.667	5.552	.707	5.885
.668	5.560	.708	5.894
.669	5.568	.709	5.902
.670	5.577	.710	5.910
.671	5.585	.711	5.918
.672	5.593	.712	5.927
.673	5.602	.713	5.935
.674	5.610	.714	5.944
.675	5.618	.715	5.952
.676	5.627	.716	5.960
.677	5.635	.717	5.968
.678	5.643	.718	5.977
.679	5.652	.719	5.985
.680	5.660	.720	5.994
.681	5.668	.721	6.002
.682	5.677	.722	6.010
.683	5.685	.723	6.018
.684	5.693	.724	6.027

Specific Gravity	Pounds per U.S. Gallon	Specific Gravity	Pounds per U.S. Gallon	Specific Gravity	Pounds per U.S. Gallon	Specific Gravity	Pounds per U.S. Gallon
.725	6.035	.765	6.369	.805	6.702	.845	7.036
.726	6.044	.766	6.377	.806	6.711	.846	7.044
.727	6.052	.767	6.386	.807	6.719	.847	7.052
.728	6.060	.768	6.394	.808	6.727	.848	7.061
.729	6.068	.769	6.402	.809	6.736	.849	7.069
.730	6.077	.770	6.410	.810	6.744	.850	7.078
.731	6,085	.771	6.419	.811	6.752	.851	7.086
.732	6.094	.772	6.427	.812	6.761	.852	7.094
.733	6.102	.773	6.436	.813	6.769	.853	7.103
.734	6.110	.774	6.444	.814	6.777	.854	7.111
.735	6.119	.775	6.452	.815	6.786	.855	7.119
.736	6.127	.776	6.460	.816	6.794		
.737	6.135	.777	6.469	.817	6.802		
.738	6.144	.778	6.477	.818	6.811		
.739	6.152	.779	6.486	.819	6.819		
.740	6.160	.780	6.494	.820	6.827		
.741	6.169	.781	6.502	.821	6.836		
.742	6.177	.782	6.510	.822	6.844		
.743	6.185	.783	6.519	.823	6.852		
.744	6.194	.784	6.527	.824	6.861		
.745	6.202	.785	6.536	.825	6.869		
.746	6.210	.786	6.544	.826	6.877		
.747	6.219	.787	6.552	.827	6.886		
.748	6.227	.788	6.560	.828	6.894		
.749	6.235	.789	6.569	.829	6.902		
.750	6.244	.790	6.577	.830	6.911		
.751	6.252	.791	6.586	.831	6.919		
.752	6.260	.792	6.594	.832	6.927		
.753	6.269	.793	6.602	.833	6.936		
.754	6.277	.794	6.611	.834	6.944		
.755	6.285	.795	6.619	.835	6.952		
.756	6.294	.796	6.627	.836	6.961		
.757	6.302	.797	6.636	.837	6.969		
.758	6.310	.798	6.644	.838	6.978		
,759	6,319	.799	6.652	.839	6.986		
.760	6.327	.800	6.661	.840	6.994		
.761	6.335	.801	6.669	.841	7.002		
.762	6.344	.802	6.677	.842	7.011		
.763	6.352	.803	6.686	.843	7.019		
.764	6.360	.804	6.694	.844	7.028		

SUPPLEMENTAL WEIGHT AND BALANCE DATE AND EQUIPMENT LIST

MAKE_____   SERIAL NO._____

MODEL_____   REG. NO._____

PREPARED BY_____   DATE_____

ITEM DESCRIPTION	WEIGHT	ARM	MOMENT

CATEGORY	EMPTY WEIGHT	EMPTY CENTER OF GRAVITY	USEFUL LOAD

PASSENGER AND CARGO MANIFEST

PASSENGER - CARGO LIST	WEIGHT	SEAT COMPT	INDEX
PASSENGER - CARGO - MAIN CABIN TOTAL			

FUEL				
WEIGHT TANK INDEX				
R-MAIN	BAGGAGE COMPT		NOSE	
L-MAIN	BAGGAGE COMPT		REAR	
R-AUX	PILOT - COPILOT			
L-AUX	OIL			
NOSE	FUEL		TOTAL	
TOTAL	EMPTY WEIGHT			
	TAKEOFF WEIGHT			

AIRCRAFT _____

FLIGHT _____

ROUTE _____

DATE _____

I HEREBY CERTIFY THAT THE ABOVE
TAKEOFF WEIGHT AND INDEX IS
WITHIN LIMITS AS SPECIFIED IN
THE FLIGHT OPERATIONS MANUAL.

_____

PILOT IN COMMAND

WEIGHT AND BALANCE REPORT

LICENSE NO. ————————————— MODEL ————————— SERIAL NO. —————

PREPARED BY —————————————————

LICENSE NO. —————————————————

DATE ————————————— DATUM LINE —————————————

LOCATION MAIN WHEELS————————— DISTANCE C.L. WHEELS TO NOSE

TAIL WHEEL —————————————

Aircraft weighted with oil in tank ————————————— gallons at —————————

LEVELING MEANS —————————————————

REACTION	GROSS	TARE	NET
Left Wheel	———	———	———
Right Wheel	———	———	———
*Nose Wheel	———	———	———
		Total	———

C.G. LOCATION AFT OF DATUM = —————————————————

REMOVAL OF OIL

	WEIGHT	ARM	MOMENT
As weighed			
Oil			
Empty Weight			

EMPTY WEIGHT C.G. = —————————

INVESTIGATION OF CRITICAL C.G. MOVEMENT

ITEM	Most Forward C.G. WEIGHT	ARM	MOMENT	Most Rearward C.G. WEIGHT	ARM	MOMENT
Airplane as weighed						
Pilot						
Passengers						
Oil						
Fuel						
Baggage						
Others - Aux. Fuel Tank						

TOTAL APPROVED C.G. LIMITS ARE ————————— AND —————————

THE USEFUL LOAD =   Approved Gross Wt.   Empty Wt.

(-   ) = Normal Category

(-   ) = Utility Category

* Strike out work not applicable

88

FINAL EXAM

Aircraft Weight and Balance

Student_____

Grade _____

Place a circle around the letter for the correct answer to each of the following questions.

Using the Weight and Balance Data for the Citation II, answer the following five question. This data will be found on the preceding pages.

NOTE: All weights are taken from the jackpoints.

1. What is the Empty Weight?

   A. 7100

   B. 7200

   C. 7252.8

2. What is the Empty CG?

   A. 300.2

   B. 298

   C. 300.1

3. What is the Basic Weight?

   A. 7200

   B. 7252.8

   C. 7300

4. What is the Basic Moment?

   A. 2276565.0

   B. 2176565.2

   C. 23565.2

5. Using the scale weights from the Empty CG problem, determine the Lateral CG. The distance between the butt line to the gear is 121.7 inches. What is the Lateral CG in inches?

   A. 1.69

   B. −1.69

   C. 1.09

Using the loading charts and tables, answer the following questions. **NOTE:** Use the nine place configuration.

6. What is the total payload?

   A. 1400

   B. 1442

   C. 2810

7. What is the payload moment?

   A. 2810.2

   B. 2744.8

   C. 2808.4

8. What is the Zero Fuel Weight?

   A. 8600.8

   B. 8652.8

   C. 8562.8

9. What is the Ramp Weight?

   A. 13500.8

   B. 13440.8

   C. 13452.8

10. What is the Takeoff Weight?

    A. 13300.8

    B. 13252.8

    C. 13362.8

11. What is the MAC at Takeoff?

    A. 21% to 22%

    B. 23% to 24%

    C. 25% to 26%

    NOTE: Use tables.

12. What is the Landing Weight?

    A. 10252

    B. 29150

    C. 12700

13. Is the aircraft still within MAC at landing?

    A. Yes

    B. No

14. When would the movement of passengers be most critical?

    A. Shortly after takeoff.

    B. Shortly before landing.

15. Which seat in the aircraft will have the most effect on movement to the aft?

    A. Seats 1 and 2

    B. Seat 1

    C. Seat 6

16. What is the Moment Change for Seat 1?

    A. 329.8

    B. 249.2

    C. 310.4

As optional equipment, the following components will be added to the aircraft: De-icer Boot for the Stabilizer, 6.38 pounds, Arm 501.58; De-icer Boot for the Fin, 2.09 pounds, Arm 528.19; and a Magazine Rack, 1.75 pounds, Arm 306.25.

17. What is the New Basic Weight?

    A. 7252.8

    B. 10.22

    C. 7263.02

18. What is the New Basic Moment?

    A. 21764.6

    B. 22764.6

    C. 22200.0

90

19. The standard 19 cell battery of 80 pounds will be replaced with an optional battery of 83 pounds. Both are located at Arm 366.72. What will be the new moment?

    A. 1100.16

    B. 22090

    C. 22310

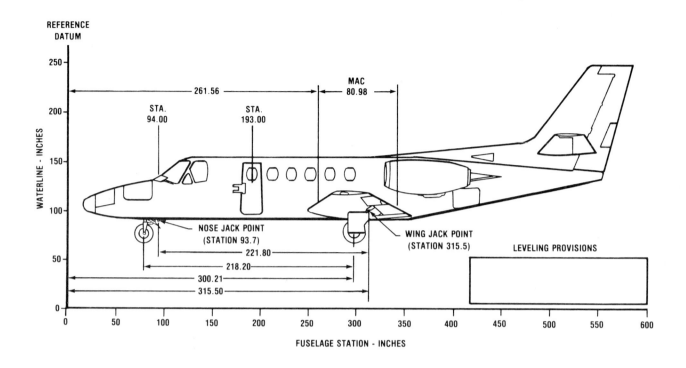

REFERENCE DATUM

250

200

WATERLINE - INCHES

150

100

50

0

261.56

MAC
80.98

STA.
94.00

STA.
193.00

NOSE JACK POINT
(STATION 93.7)

WING JACK POINT
(STATION 315.5)

LEVELING PROVISIONS

221.80

218.20

300.21

315.50

0   50   100   150   200   250   300   350   400   450   500   550   600

FUSELAGE STATION - INCHES

AIRPLANE AS WEIGHED TABLE

POSITION	SCALE READING	SCALE DRIFT	TARE	NET WEIGHT
LEFT WING	3402	2	0	
RIGHT WING	3302	2	0	
NOSE	500	0	0	
AIRPLANE TOTAL AS WEIGHED				

FUSELAGE STATION OF AFT WEIGHING POINT

IF WEIGHED ON JACK POINTS USE 221.80
IF WEIGHED ON WHEELS USE 218.20

CG ARM OF AIRPLANE AS WEIGHED USING JACK POINTS OR WHEELS* = (   ) - ($\frac{(\quad) \times (\quad)}{(\quad)}$) = (   )

NOTE

IT IS THE RESPONSIBILITY OF THE OPERATOR TO INSURE THAT THE AIRPLANE IS LOADED PROPERLY.

BASIC EMPTY WEIGHT AND CENTER OF GRAVITY TABLE

ITEM		WEIGHT - POUNDS	CG ARM - INCHES	MOMENT (INCH-POUNDS/100)
△ AIRPLANE (CALCULATED OR AS WEIGHED)				
DRAINABLE UNUSABLE FUEL AT 6.75 POUNDS PER GALLON LEFT AND RIGHT WING		52.8	298.4	157.6
BASIC EMPTY WEIGHT				

WEIGHT AND MOMENT TABLES
CITATION II STANDARD
MODEL   550/551

CREW AND PASSENGER SEATS

WEIGHT (POUNDS)	1ST OR 2ND SEATS ARM - 131"	5TH OR 6TH SEATS ARM - 232"	7TH OR 8TH SEATS ARM - 288"	9TH OR 10TH SEATS ARM - 170"	LOUNGE SEAT ARM - 167"	AFT PORTABLE SEAT ARM - 333"	THREE PLACE SIDE FACING COUCH ARM - 216"	ARM - 235"	ARM - 254"
				MOMENT/100					
50	65.5	116.0	144.0	85.0	83.5	166.5	108.0	117.5	127.0
60	78.6	139.2	172.8	102.0	100.2	199.8	129.6	141.0	152.4
70	91.7	162.4	201.6	119.0	116.9	233.1	151.2	164.5	177.8
80	104.8	185.6	230.4	136.0	133.6	266.4	172.8	188.0	203.2
90	117.9	208.8	259.2	153.0	150.3	299.7	194.4	211.5	228.6
100	131.0	232.0	288.0	170.0	167.0	333.0	216.0	235.0	254.0
110	144.1	255.2	316.8	187.0	183.7	366.3	237.6	258.5	279.4
120	157.2	278.4	345.6	204.0	200.4	399.6	259.2	282.0	304.8
130	170.3	301.6	374.4	221.0	217.1	432.9	280.8	305.5	330.2
140	183.4	324.8	403.2	238.0	233.8	466.2	302.4	329.0	355.6
150	196.5	348.0	432.0	255.0	250.5	499.5	324.0	352.5	381.0
160	209.6	371.2	460.8	272.0	267.2	532.8	345.6	376.0	406.4
170	222.7	394.4	489.6	289.0	283.9	566.1	367.2	399.5	431.8
180	235.8	417.6	518.4	306.0					
190	248.9	440.8	547.2	323.0					
200	262.0	464.0	576.0	340.0					
210	275.1	487.2	604.8	357.0					
220	288.2	510.4	633.6	374.0					
230	301.3	533.6	662.4	391.0					
240	314.4	556.8	691.2	408.0					
250	327.5	580.0	720.0	425.0					
260	340.6	603.2	748.8	442.0					
270	353.7	626.4	777.6	459.0					
280	366.8	649.6	806.4	476.0					
290	379.9	672.8	835.2	493.0					
300	393.0	696.0	864.0	510.0					
310	406.1	719.2	892.8	527.0					
320	419.2	742.4	921.6	544.0					
330	432.3	765.6	950.4	561.0					
340	445.4	788.8	979.2	578.0					

CABINET CONTENTS

WEIGHT (POUNDS)	MIDSHIP REFRESHMENT CENTER ARM - 213"	CABINET WITH MAPCO ARM - 213"	3 DRAWER CABINET ARM - 213"	STEREO CABINET ARM - 213"
		MOMENT/100		
5	10.6	10.6	10.6	10.6
10	21.3	21.3	21.3	21.3
15	32.0	32.0	32.0	32.0
20			42.6	

BAGGAGE CONTENTS

WEIGHT (POUNDS)	NOSE COMPARTMENT ARM - 74"	AFT CABIN COMPARTMENT ARM - 321"	ARM - 338"	TAIL CONE COMPARTMENT ARM - 442"
		MOMENT/100		
20	14.8	64.2	67.6	88.4
40	29.6	128.4	135.2	176.8
60	44.4	192.6	202.8	265.2
80	59.2	256.8	270.4	353.6
100	74.0	321.0	338.0	442.0
120	88.8	385.2	405.6	530.4
140	103.6	449.4	473.2	618.8
160	118.4	513.6	540.8	707.2
180	133.2	577.8	608.4	795.6
200	148.0	642.0	676.0	884.0
220	162.8	706.2		
240	177.6	770.4		
260	192.4	834.6		
280	207.2	898.8		
300	222.0	963.0		
320	236.8	1027.2		
340	251.6	1091.4		
350	259.0	1123.5		
360		1155.6		
380		1219.8		
400		1284.0		

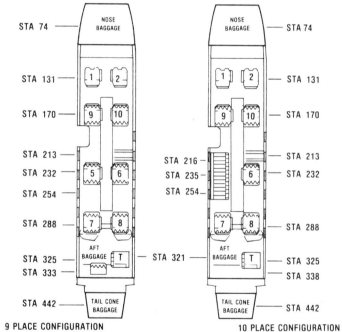

CREW AND PASSENGER SEATS AND BAGGAGE MOMENT ARMS

9 PLACE CONFIGURATION   10 PLACE CONFIGURATION

WEIGHT (POUNDS)	MOMENT/100 ARM VARIES	WEIGHT (POUNDS)	MOMENT/100 ARM VARIES
100	298	2600	7415
200	590	2700	7700
300	879	2800	7984
400	1164	2900	8269
500	1449	3000	8555
600	1732	3100	8840
700	2015	3200	9125
800	2298	3300	9412
900	2582	3400	9697
1000	2866	3500	9982
1100	3150	3600	10269
1200	3434	3700	10556
1300	3719	3800	10843
1400	4003	3900	11131
1500	4289	4000	11418
1600	4573	4100	11706
1700	4857	4200	11993
1800	5141	4300	12281
1900	5426	4400	12568
2000	5710	4500	12856
2100	5994	4600	13142
2200	6279	4700	13430
2300	6563	4800	13721
2400	6847	4900	14008
2500	7131	5008	14319

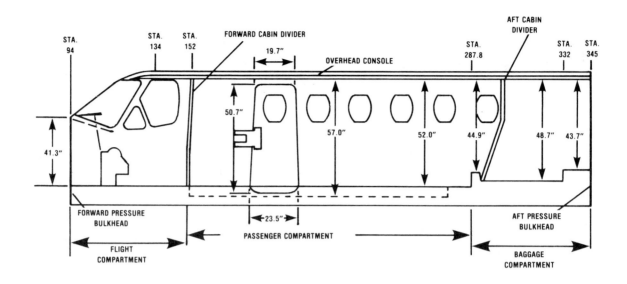

PERSONNEL MOVEMENT TO AFT TOILET						
OCCUPANT WEIGHT (POUNDS)	SEAT 1 OR 2	SEAT 3 OR 4 OPT 1	SEAT 5 OR 6 STD-OPT 1		SEAT 7 OR 8 STD-OPT 1	SEAT 9 OR 10 STD-OPT 1
	MOMENT/100					
50	+97.0	+54.0	+46.5	+36.0	+18.5	+77.5
60	+116.4	+64.8	+55.8	+43.2	+22.2	+93.0
70	+135.8	+75.6	+65.1	+50.4	+25.9	+108.5
80	+155.2	+86.4	+74.4	+57.6	+29.6	+124.0
90	+174.6	+97.2	+83.7	+64.8	+33.3	+139.5
100	+194.0	+108.0	+93.0	+72.0	+37.0	+155.0
110	+213.4	+118.8	+102.3	+79.2	+40.7	+170.5
120	+232.8	+129.6	+111.6	+86.4	+44.4	+186.0
130	+252.2	+140.4	+120.9	+93.6	+48.1	+201.5
140	+271.6	+151.2	+130.2	+100.8	+51.8	+217.0
150	+291.0	+162.0	+139.5	+108.0	+55.5	+232.5
160	+310.4	+172.8	+148.8	+115.2	+59.2	+248.0
170	+329.8	+183.6	+158.1	+122.4	+62.9	+263.5
180	+349.2	+194.4	+167.4	+129.6	+66.6	+279.0
190	+368.6	+205.2	+176.7	+136.8	+70.3	+294.5
200	+388.0	+216.0	+186.0	+144.0	+74.0	+310.0
210	+407.4	+226.8	+195.3	+151.2	+77.7	+325.5
220	+426.8	+237.6	+204.6	+158.4	+81.4	+341.0
230	+446.2	+248.4	+213.9	+165.6	+85.1	+456.5
240	+465.6	+259.2	+223.2	+172.8	+88.8	+372.0
250	+485.0	+270.0	+232.5	+180.0	+92.5	+387.5
260	+504.4	+280.8	+214.8	+187.2	+96.2	+403.0
270	+523.8	+291.6	+251.1	+194.4	+99.9	+418.5
280	+543.2	+302.4	+260.4	+201.6	+103.6	+434.0
290	+562.6	+313.1	+269.7	+208.8	+107.3	+449.5
300	+582.0	+324.0	+279.0	+216.0	+111.0	+465.0

Crew and Passenger Movement Table
Table 7

WEIGHT AND BALANCE FORM
CITATION II
Model 550

PAYLOAD COMPUTATIONS				R E F	ITEM	WEIGHT	MOMENT/ 100
ITEM OCCUPANTS OR CARGO	ARM	WEIGHT	MOMENT/ 100	1.	BASIC EMPTY WEIGHT		
SEAT 1		170		2.	PAYLOAD		
SEAT 2		160		3.	ZERO FUEL WEIGHT (sub-total) (Do not exceed maximum zero fuel weight) 9500 Std 11,000 Opt		
SEAT 9		170					
SEAT 10		150					
SEAT 5		170					
SEAT 6		130		4.	FUEL LOADING	4800	
SEAT 7		130		5.	RAMP WEIGHT (sub-total) (Do not exceed maximum ramp weight of 13,500 pounds)		
SEAT 8		200					
SEAT __							
SEAT __				6.	LESS FUEL FOR TAXIING	200	
SEAT (AFT DIVAN)				7.	TAKEOFF WEIGHT (Do not exceed maximum takeoff weight of 13,300 pounds)		
SIDE FACING COUCH							
AFT PORTABLE SEAT				8.	LESS FUEL TO DESTINATION	3000	
TOILET				9.	LANDING WEIGHT (Do not exceed maximum landing weight of 12,700 pounds)		
BAGGAGE							
NOSE		80					
FWD CABIN							
AFT CABIN		40					
TAIL CONE							
CABINET							
MIDSHIP							
AFT REFRESHMENT CENTER							
PAYLOAD							

Total must be within approved weight and C.G. limits. It is the responsibility of the operator to insure that the airplane is loaded properly. The Basic Empty Weight C.G. is noted on the Airplane Weighing Form. If the airplane has been altered, refer to the weight and balance record for this information.

GROSS WEIGHT POUNDS	18% MAC 276.1	19% MAC 276.9	20% MAC 277.8	21% MAC 278.6	22% MAC 279.4	23% MAC 280.2	24% MAC 281.0	25% MAC 281.8	26% MAC 282.6	27% MAC 283.4	28% MAC 284.2	29% MAC 285.0	30% MAC 285.8
9800.			27220.	27299.	27379.	27458.	27537.	27617.	27696.	27775.	27855.	27934.	28014.
9810.			27248.	27327.	27407.	27486.	27565.	27645.	27724.	27804.	27883.	27963.	28042.
9820.			27275.	27355.	27434.	27514.	27594.	27673.	27753.	27832.	27912.	27991.	28071.
9830.			27303.	27383.	27462.	27542.	27622.	27701.	27781.	27860.	27940.	28020.	28099.
9840.			27331.	27411.	27490.	27570.	27650.	27729.	27809.	27889.	27968.	28048.	28128.
9850.			27359.	27439.	27518.	27598.	27678.	27758.	27837.	27917.	27997.	28077.	28156.
9860.			27387.	27466.	27546.	27626.	27706.	27786.	27866.	27945.	28025.	28105.	28185.
9870.			27414.	27494.	27574.	27654.	27734.	27814.	27894.	27974.	28054.	28134.	28214.
9880.			27442.	27522.	27602.	27682.	27762.	27842.	27922.	28002.	28082.	28162.	28242.
9890.			27470.	27550.	27630.	27710.	27790.	27870.	27950.	28030.	28111.	28191.	28271.
9900.			27498.	27578.	27658.	27738.	27818.	27898.	27979.	28059.	28139.	28219.	28299.
9910.			27525.	27606.	27686.	27766.	27846.	27927.	28007.	28087.	28167.	28248.	28328.
9920.			27553.	27634.	27714.	27794.	27875.	27955.	28035.	28116.	28196.	28276.	28357.
9930.			27581.	27661.	27742.	27822.	27903.	27983.	28063.	28144.	28224.	28305.	28385.
9940.			27609.	27689.	27770.	27850.	27931.	28011.	28092.	28172.	28253.	28333.	28414.
9950.			27637.	27717.	27798.	27878.	27959.	28039.	28120.	28201.	28281.	28362.	28442.
9960.			27664.	27745.	27826.	27906.	27987.	28068.	28148.	28229.	28310.	28390.	28471.
9970.			27692.	27773.	27854.	27934.	28015.	28096.	28176.	28257.	28338.	28419.	28499.
9980.			27720.	27801.	27881.	27962.	28043.	28124.	28205.	28286.	28366.	28447.	28528.
9990.			27748.	27829.	27909.	27990.	28071.	28152.	28233.	28314.	28395.	28476.	28557.
10000.			27775.	27856.	27937.	28018.	28099.	28180.	28261.	28342.	28423.	28504.	28585.
10010.			27803.	27884.	27965.	28046.	28127.	28208.	28290.	28371.	28452.	28533.	28614.
10020.			27831.	27912.	27993.	28074.	28156.	28237.	28318.	28399.	28480.	28561.	28642.
10030.			27859.	27940.	28021.	28102.	28184.	28265.	28346.	28427.	28509.	28590.	28671.
10040.			27886.	27968.	28049.	28130.	28212.	28293.	28374.	28456.	28537.	28618.	28700.
10050.			27914.	27996.	28077.	28158.	28240.	28321.	28403.	28484.	28565.	28647.	28728.
10060.			27942.	28024.	28105.	28186.	28268.	28349.	28431.	28512.	28594.	28675.	28757.
10070.			27970.	28051.	28133.	28214.	28296.	28378.	28459.	28541.	28622.	28704.	28785.
10080.			27998.	28079.	28161.	28242.	28324.	28406.	28487.	28569.	28651.	28732.	28814.
10090.			28025.	28107.	28189.	28270.	28352.	28434.	28516.	28597.	28679.	28761.	28842.
10100.			28053.	28135.	28217.	28299.	28380.	28462.	28544.	28626.	28707.	28789.	28871.
10110.			28081.	28163.	28245.	28327.	28408.	28490.	28572.	28654.	28736.	28818.	28900.
10120.			28109.	28191.	28273.	28355.	28437.	28518.	28600.	28682.	28764.	28846.	28928.
10130.			28136.	28219.	28301.	28383.	28465.	28547.	28629.	28711.	28793.	28875.	28957.
10140.			28164.	28246.	28328.	28411.	28493.	28575.	28657.	28739.	28821.	28903.	28985.
10150.			28192.	28274.	28356.	28439.	28521.	28603.	28685.	28767.	28850.	28932.	29014.
10160.			28220.	28302.	28384.	28467.	28549.	28631.	28713.	28796.	28878.	28960.	29043.
10170.			28248.	28330.	28412.	28495.	28577.	28659.	28742.	28824.	28906.	28989.	29071.
10180.			28275.	28358.	28440.	28523.	28605.	28688.	28770.	28852.	28935.	29017.	29100.
10190.			28303.	28386.	28468.	28551.	28633.	28716.	28798.	28881.	28963.	29046.	29128.
10200.			28331.	28413.	28496.	28579.	28661.	28744.	28827.	28909.	28992.	29074.	29157.
10210.			28359.	28441.	28524.	28607.	28689.	28772.	28855.	28937.	29020.	29103.	29185.
10220.			28386.	28469.	28552.	28635.	28718.	28800.	28883.	28966.	29049.	29131.	29214.
10230.			28414.	28497.	28580.	28663.	28746.	28828.	28911.	28994.	29077.	29160.	29243.
10240.			28442.	28525.	28608.	28691.	28774.	28857.	28940.	29022.	29105.	29188.	29271.
10250.			28470.	28553.	28636.	28719.	28802.	28885.	28968.	29051.	29134.	29217.	29300.
10260.			28498.	28581.	28664.	28747.	28830.	28913.	28996.	29079.	29162.	29245.	29328.
10270.			28525.	28608.	28692.	28775.	28858.	28941.	29024.	29108.	29191.	29274.	29357.
10280.			28553.	28636.	28720.	28803.	28886.	28969.	29053.	29136.	29219.	29302.	29386.
10290.			28581.	28664.	28748.	28831.	28914.	28998.	29081.	29164.	29248.	29331.	29414.
10300.			28609.	28692.	28775.	28859.	28942.	29026.	29109.	29193.	29276.	29359.	29443.
10310.			28636.	28720.	28803.	28887.	28970.	29054.	29137.	29221.	29304.	29388.	29471.
10320.			28664.	28748.	28831.	28915.	28998.	29082.	29166.	29249.	29333.	29416.	29500.
10330.			28692.	28776.	28859.	28943.	29027.	29110.	29194.	29278.	29361.	29445.	29529.
10340.			28720.	28803.	28887.	28971.	29055.	29138.	29222.	29306.	29390.	29473.	29557.
10350.			28748.	28831.	28915.	28999.	29083.	29167.	29250.	29334.	29418.	29502.	29586.
10360.			28775.	28859.	28943.	29027.	29111.	29195.	29279.	29363.	29446.	29530.	29614.
10370.			28803.	28887.	28971.	29055.	29139.	29223.	29307.	29391.	29475.	29559.	29643.
10380.			28831.	28915.	28999.	29083.	29167.	29251.	29335.	29419.	29503.	29587.	29671.
10390.			28859.	28943.	29027.	29111.	29195.	29279.	29363.	29448.	29532.	29616.	29700.
10400.			28886.	28971.	29055.	29139.	29223.	29308.	29392.	29476.	29560.	29644.	29729.
10410.			28914.	28998.	29083.	29167.	29251.	29336.	29420.	29504.	29589.	29673.	29757.
10420.			28942.	29026.	29111.	29195.	29279.	29364.	29448.	29533.	29617.	29701.	29786.
10430.			28970.	29054.	29139.	29223.	29308.	29393.	29477.	29561.	29645.	29730.	29814.
10440.			28998.	29082.	29167.	29251.	29336.	29420.	29505.	29589.	29674.	29758.	29843.
10450.			29025.	29110.	29195.	29279.	29364.	29448.	29533.	29618.	29702.	29787.	29872.
10460.			29053.	29138.	29222.	29307.	29392.	29477.	29561.	29646.	29731.	29815.	29900.
10470.			29081.	29166.	29250.	29335.	29420.	29505.	29590.	29674.	29759.	29844.	29929.
10480.			29109.	29193.	29278.	29363.	29448.	29533.	29618.	29703.	29788.	29872.	29957.
10490.			29136.	29221.	29306.	29391.	29476.	29561.	29646.	29731.	29816.	29901.	29986.

96

GROSS WEIGHT POUNDS	18% MAC 276.1	19% MAC 276.9	20% MAC 277.8	21% MAC 278.6	22% MAC 279.4	23% MAC 280.2	24% MAC 281.0	25% MAC 281.8	26% MAC 282.6	27% MAC 283.4	28% MAC 284.2	29% MAC 285.0	30% MAC 285.8
							MOMENT/100						
10500.			29164.	29249.	29334.	29419.	29504.	29589.	29674.	29759.	29844.	29929.	30014.
10510.			29192.	29277.	29362.	29447.	29532.	29617.	29703.	29788.	29873.	29958.	30043.
10520.			29220.	29305.	29390.	29475.	29560.	29646.	29731.	29816.	29901.	29986.	30072.
10530.			29247.	29333.	29418.	29503.	29589.	29674.	29759.	29844.	29930.	30015.	30100.
10540.			29275.	29361.	29446.	29531.	29617.	29702.	29787.	29873.	29958.	30043.	30129.
10550.			29303.	29388.	29474.	29559.	29645.	29730.	29816.	29901.	29987.	30072.	30157.
10560.			29331.	29416.	29502.	29587.	29673.	29758.	29844.	29929.	30015.	30100.	30186.
10570.			29359.	29444.	29530.	29615.	29701.	29787.	29872.	29958.	30043.	30129.	30215.
10580.			29386.	29472.	29558.	29643.	29729.	29815.	29900.	29986.	30072.	30157.	30243.
10590.			29414.	29500.	29586.	29671.	29757.	29843.	29929.	30014.	30100.	30186.	30272.
10600.			29442.	29528.	29614.	29699.	29785.	29871.	29957.	30043.	30129.	30214.	30300.
10610.			29470.	29556.	29642.	29727.	29813.	29899.	29985.	30071.	30157.	30243.	30329.
10620.			29497.	29583.	29669.	29755.	29841.	29927.	30013.	30099.	30185.	30271.	30357.
10630.				29611.	29697.	29783.	29870.	29956.	30042.	30128.	30214.	30300.	30386.
10640.				29639.	29725.	29812.	29898.	29984.	30070.	30156.	30242.	30328.	30415.
10650.				29667.	29753.	29840.	29926.	30012.	30098.	30185.	30271.	30357.	30443.
10660.				29695.	29781.	29868.	29954.	30040.	30127.	30213.	30299.	30386.	30472.
10670.				29723.	29809.	29896.	29982.	30068.	30155.	30241.	30328.	30414.	30500.
10680.				29751.	29837.	29924.	30010.	30097.	30183.	30270.	30356.	30443.	30529.
10690.				29778.	29865.	29952.	30038.	30125.	30211.	30298.	30384.	30471.	30558.
10700.				29806.	29893.	29980.	30066.	30153.	30240.	30326.	30413.	30500.	30586.
10710.				29834.	29921.	30008.	30094.	30181.	30268.	30355.	30441.	30528.	30615.
10720.				29862.	29949.	30036.	30122.	30209.	30296.	30383.	30470.	30557.	30643.
10730.				29890.	29977.	30064.	30151.	30237.	30324.	30411.	30498.	30585.	30672.
10740.				29918.	30005.	30092.	30179.	30266.	30353.	30440.	30527.	30614.	30701.
10750.				29946.	30033.	30120.	30207.	30294.	30381.	30468.	30555.	30642.	30729.
10760.				29973.	30061.	30148.	30235.	30322.	30409.	30496.	30583.	30671.	30758.
10770.				30001.	30089.	30176.	30263.	30350.	30437.	30525.	30612.	30699.	30786.
10780.				30029.	30116.	30204.	30291.	30378.	30466.	30553.	30640.	30728.	30815.
10790.				30057.	30144.	30232.	30319.	30407.	30494.	30581.	30669.	30756.	30843.
10800.				30085.	30172.	30260.	30347.	30435.	30522.	30610.	30697.	30785.	30872.
10810.				30113.	30200.	30288.	30375.	30463.	30550.	30638.	30726.	30813.	30901.
10820.				30141.	30228.	30316.	30403.	30491.	30579.	30666.	30754.	30842.	30929.
10830.				30168.	30256.	30344.	30432.	30519.	30607.	30695.	30782.	30870.	30958.
10840.				30196.	30284.	30372.	30460.	30547.	30635.	30723.	30811.	30899.	30986.
10850.				30224.	30312.	30400.	30488.	30576.	30663.	30751.	30839.	30927.	31015.
10860.				30252.	30340.	30428.	30516.	30604.	30692.	30780.	30868.	30956.	31044.
10870.				30280.	30368.	30456.	30544.	30632.	30720.	30808.	30896.	30984.	31072.
10880.				30308.	30396.	30484.	30572.	30660.	30748.	30836.	30924.	31013.	31101.
10890.				30336.	30424.	30512.	30600.	30688.	30777.	30865.	30953.	31041.	31129.
10900.				30363.	30452.	30540.	30628.	30717.	30805.	30893.	30981.	31070.	31158.
10910.				30391.	30480.	30568.	30656.	30745.	30833.	30921.	31010.	31098.	31186.
10920.				30419.	30508.	30596.	30684.	30773.	30861.	30950.	31038.	31127.	31215.
10930.				30447.	30536.	30624.	30713.	30801.	30890.	30978.	31067.	31155.	31244.
10940.				30475.	30563.	30652.	30741.	30829.	30918.	31006.	31095.	31184.	31272.
10950.				30503.	30591.	30680.	30769.	30857.	30946.	31035.	31123.	31212.	31301.
10960.				30531.	30619.	30708.	30797.	30886.	30974.	31063.	31152.	31241.	31329.
10970.				30558.	30647.	30736.	30825.	30914.	31003.	31091.	31180.	31269.	31358.
10980.				30586.	30675.	30764.	30853.	30942.	31031.	31120.	31209.	31298.	31387.
10990.				30614.	30703.	30792.	30881.	30970.	31059.	31148.	31237.	31326.	31415.
11000.				30642.	30731.	30820.	30909.	30998.	31087.	31176.	31266.	31355.	31444.
11010.				30670.	30759.	30848.	30937.	31027.	31116.	31205.	31294.	31383.	31472.
11020.				30698.	30787.	30876.	30965.	31055.	31144.	31233.	31322.	31412.	31501.
11030.				30726.	30815.	30904.	30994.	31083.	31172.	31262.	31351.	31440.	31529.
11040.				30753.	30843.	30932.	31022.	31111.	31200.	31290.	31379.	31469.	31558.
11050.				30781.	30871.	30960.	31050.	31139.	31229.	31318.	31408.	31497.	31587.
11060.				30809.	30899.	30988.	31078.	31167.	31257.	31347.	31436.	31526.	31615.
11070.				30837.	30927.	31016.	31106.	31196.	31285.	31375.	31465.	31554.	31644.
11080.				30865.	30955.	31044.	31134.	31224.	31313.	31403.	31493.	31583.	31672.
11090.				30893.	30983.	31072.	31162.	31252.	31342.	31432.	31521.	31611.	31701.
11100.				30921.	31010.	31100.	31190.	31280.	31370.	31460.	31550.	31640.	31730.
11110.				30948.	31038.	31128.	31218.	31308.	31398.	31488.	31578.	31668.	31758.
11120.				30976.	31066.	31156.	31246.	31336.	31427.	31517.	31607.	31697.	31787.
11130.				31004.	31094.	31184.	31275.	31365.	31455.	31545.	31635.	31725.	31815.
11140.				31032.	31122.	31212.	31303.	31393.	31483.	31573.	31663.	31754.	31844.
11150.				31060.	31150.	31240.	31331.	31421.	31511.	31602.	31692.	31782.	31873.
11160.				31088.	31178.	31268.	31359.	31449.	31540.	31630.	31720.	31811.	31901.
11170.				31116.	31206.	31296.	31387.	31477.	31568.	31658.	31749.	31839.	31930.
11180.				31143.	31234.	31324.	31415.	31506.	31596.	31687.	31777.	31868.	31958.
11190.				31171.	31262.	31353.	31443.	31534.	31624.	31715.	31806.	31896.	31987.

GROSS WEIGHT POUNDS	MOMENT/100												
	18% MAC 276.1	19% MAC 276.9	20% MAC 277.8	21% MAC 278.6	22% MAC 279.4	23% MAC 280.2	24% MAC 281.0	25% MAC 281.8	26% MAC 282.6	27% MAC 283.4	28% MAC 284.2	29% MAC 285.0	30% MAC 285.8
11200.				31199.	31290.	31381.	31471.	31562.	31653.	31743.	31834.	31925.	32015.
11210.				31227.	31318.	31409.	31499.	31590.	31681.	31772.	31862.	31953.	32044.
11220.				31255.	31346.	31437.	31527.	31618.	31709.	31800.	31891.	31982.	32073.
11230.				31283.	31374.	31465.	31556.	31646.	31737.	31828.	31919.	32010.	32101.
11240.				31311.	31402.	31493.	31584.	31675.	31766.	31857.	31948.	32039.	32130.
11250.				31338.	31430.	31521.	31612.	31703.	31794.	31885.	31976.	32067.	32158.
11260.				31366.	31457.	31549.	31640.	31731.	31822.	31913.	32005.	32096.	32187.
11270.				31394.	31485.	31577.	31668.	31759.	31850.	31942.	32033.	32124.	32216.
11280.				31422.	31513.	31605.	31696.	31787.	31879.	31970.	32061.	32153.	32244.
11290.				31450.	31541.	31633.	31724.	31816.	31907.	31998.	32090.	32181.	32273.
11300.				31478.	31569.	31661.	31752.	31844.	31935.	32027.	32118.	32210.	32301.
11310.				31506.	31597.	31689.	31780.	31872.	31964.	32055.	32147.	32238.	32330.
11320.				31533.	31625.	31717.	31808.	31900.	31992.	32083.	32175.	32267.	32358.
11330.				31561.	31653.	31745.	31837.	31928.	32020.	32112.	32204.	32295.	32387.
11340.				31589.	31681.	31773.	31865.	31956.	32048.	32140.	32232.	32324.	32416.
11350.				31617.	31709.	31801.	31893.	31985.	32077.	32168.	32260.	32352.	32444.
11360.				31645.	31737.	31829.	31921.	32013.	32105.	32197.	32289.	32381.	32473.
11370.				31673.	31765.	31857.	31949.	32041.	32133.	32225.	32317.	32409.	32501.
11380.				31701.	31793.	31885.	31977.	32069.	32161.	32253.	32346.	32438.	32530.
11390.				31728.	31821.	31913.	32005.	32097.	32190.	32282.	32374.	32466.	32559.
11400.				31756.	31849.	31941.	32033.	32126.	32218.	32310.	32402.	32495.	32587.
11410.				31784.	31877.	31969.	32061.	32154.	32246.	32339.	32431.	32523.	32616.
11420.				31812.	31904.	31997.	32089.	32182.	32274.	32367.	32459.	32552.	32644.
11430.				31840.	31932.	32025.	32118.	32210.	32303.	32395.	32488.	32580.	32673.
11440.				31868.	31960.	32053.	32146.	32238.	32331.	32424.	32516.	32609.	32701.
11450.				31896.	31988.	32081.	32174.	32266.	32359.	32452.	32545.	32637.	32730.
11460.				31923.	32016.	32109.	32202.	32295.	32387.	32480.	32573.	32666.	32759.
11470.				31951.	32044.	32137.	32230.	32323.	32416.	32509.	32601.	32694.	32787.
11480.				31979.	32072.	32165.	32258.	32351.	32444.	32537.	32630.	32723.	32816.
11490.				32007.	32100.	32193.	32286.	32379.	32472.	32565.	32658.	32751.	32844.
11500.				32035.	32128.	32221.	32314.	32407.	32500.	32594.	32687.	32780.	32873.
11510.				32063.	32156.	32249.	32342.	32436.	32529.	32622.	32715.	32808.	32902.
11520.				32091.	32184.	32277.	32370.	32464.	32557.	32650.	32744.	32837.	32930.
11530.				32118.	32212.	32305.	32399.	32492.	32585.	32679.	32772.	32865.	32959.
11540.				32146.	32240.	32333.	32427.	32520.	32614.	32707.	32800.	32894.	32987.
11550.				32174.	32268.	32361.	32455.	32548.	32642.	32735.	32829.	32922.	33016.
11560.				32202.	32296.	32389.	32483.	32576.	32670.	32764.	32857.	32951.	33044.
11570.				32230.	32324.	32417.	32511.	32605.	32698.	32792.	32886.	32979.	33073.
11580.				32258.	32351.	32445.	32539.	32633.	32727.	32820.	32914.	33008.	33102.
11590.				32286.	32379.	32473.	32567.	32661.	32755.	32849.	32943.	33036.	33130.
11600.				32313.	32407.	32501.	32595.	32689.	32783.	32877.	32971.	33065.	33159.
11610.				32341.	32435.	32529.	32623.	32717.	32811.	32905.	32999.	33093.	33197.
11620.				32369.	32463.	32557.	32651.	32746.	32840.	32934.	33028.	33122.	33216.
11630.				32397.	32491.	32585.	32680.	32774.	32868.	32962.	33056.	33150.	33245.
11640.				32425.	32519.	32613.	32708.	32802.	32896.	32990.	33085.	33179.	33270.
11650.				32453.	32547.	32641.	32736.	32830.	32924.	33019.	33113.	33207.	33302.
11660.				32481.	32575.	32669.	32764.	32858.	32953.	33047.	33142.	33236.	33330.
11670.					32603.	32697.	32792.	32886.	32981.	33075.	33170.	33264.	33359.
11680.					32631.	32725.	32820.	32915.	33009.	33104.	33198.	33293.	33388.
11690.					32659.	32753.	32848.	32943.	33037.	33132.	33227.	33321.	33416.
11700.					32687.	32781.	32876.	32971.	33066.	33160.	33255.	33350.	33445.
11710.					32715.	32809.	32904.	32999.	33094.	33189.	33284.	33378.	33473.
11720.					32743.	32837.	32932.	33027.	33122.	33217.	33312.	33407.	33502.
11730.					32771.	32866.	32960.	33055.	33150.	33245.	33340.	33435.	33530.
11740.					32798.	32894.	32989.	33084.	33179.	33274.	33369.	33464.	33559.
11750.					32826.	32922.	33017.	33112.	33207.	33302.	33397.	33492.	33588.
11760.					32854.	32950.	33045.	33140.	33235.	33330.	33426.	33521.	33616.
11770.					32882.	32978.	33073.	33168.	33264.	33359.	33454.	33549.	33645.
11780.					32910.	33006.	33101.	33196.	33292.	33387.	33483.	33578.	33673.
11790.					32938.	33034.	33129.	33225.	33320.	33416.	33511.	33606.	33702.
11800.					32966.	33062.	33157.	33253.	33348.	33444.	33539.	33635.	33731.
11810.					32994.	33090.	33185.	33281.	33377.	33472.	33568.	33663.	33759.
11820.					33022.	33118.	33213.	33309.	33405.	33501.	33596.	33692.	33788.
11830.					33050.	33146.	33241.	33337.	33433.	33529.	33625.	33720.	33816.
11840.					33078.	33174.	33270.	33365.	33461.	33557.	33653.	33749.	33845.
11850.					33106.	33202.	33298.	33394.	33490.	33586.	33682.	33778.	33873.
11860.					33134.	33230.	33326.	33422.	33518.	33614.	33710.	33806.	33902.
11870.					33162.	33258.	33354.	33450.	33546.	33642.	33738.	33835.	33931.
11880.					33190.	33286.	33382.	33478.	33574.	33671.	33767.	33863.	33959.
11890.					33218.	33314.	33410.	33506.	33603.	33699.	33795.	33892.	33988.

GROSS WEIGHT POUNDS	18% MAC 276.1	19% MAC 276.9	20% MAC 277.8	21% MAC 278.6	22% MAC 279.4	23% MAC 280.2	24% MAC 281.0	25% MAC 281.8	26% MAC 282.6	27% MAC 283.4	28% MAC 284.2	29% MAC 285.0	30% MAC 285.8
										MOMENT/100			
11900.					33245.	33342.	33438.	33535.	33631.	33727.	33824.	33920.	34016.
11910.					33273.	33370.	33466.	33563.	33659.	33756.	33852.	33949.	34045.
11920.					33301.	33398.	33494.	33591.	33687.	33784.	33881.	33977.	34074.
11930.					33329.	33426.	33522.	33619.	33716.	33812.	33909.	34006.	34102.
11940.					33357.	33454.	33551.	33647.	33744.	33841.	33937.	34034.	34131.
11950.					33385.	33482.	33579.	33675.	33772.	33869.	33966.	34063.	34159.
11960.					33413.	33510.	33607.	33704.	33800.	33897.	33994.	34091.	34188.
11970.					33441.	33538.	33635.	33732.	33829.	33926.	34023.	34120.	34216.
11980.					33469.	33566.	33663.	33760.	33857.	33954.	34051.	34148.	34245.
11990.					33497.	33594.	33691.	33788.	33885.	33982.	34079.	34177.	34274.
12000.					33525.	33622.	33719.	33816.	33914.	34011.	34108.	34205.	34302.
12010.					33553.	33650.	33747.	33845.	33942.	34039.	34136.	34234.	34331.
12020.					33581.	33678.	33775.	33873.	33970.	34067.	34165.	34262.	34359.
12030.					33609.	33706.	33803.	33901.	33998.	34096.	34193.	34291.	34388.
12040.					33637.	33734.	33832.	33929.	34027.	34124.	34222.	34319.	34417.
12050.					33665.	33762.	33860.	33957.	34055.	34152.	34250.	34348.	34445.
12060.					33692.	33790.	33888.	33985.	34083.	34181.	34278.	34376.	34474.
12070.					33720.	33818.	33916.	34014.	34111.	34209.	34307.	34405.	34502.
12080.					33748.	33846.	33944.	34042.	34140.	34237.	34335.	34433.	34531.
12090.					33776.	33874.	33972.	34070.	34168.	34266.	34364.	34462.	34560.
12100.					33804.	33902.	34000.	34098.	34196.	34294.	34392.	34490.	34588.
12110.					33832.	33930.	34028.	34126.	34224.	34322.	34421.	34519.	34617.
12120.					33860.	33958.	34056.	34155.	34253.	34351.	34449.	34547.	34645.
12130.					33888.	33986.	34084.	34183.	34281.	34379.	34477.	34576.	34674.
12140.					33916.	34014.	34113.	34211.	34309.	34408.	34506.	34604.	34702.
12150.					33944.	34042.	34141.	34239.	34337.	34436.	34534.	34633.	34731.
12160.					33972.	34070.	34169.	34267.	34366.	34464.	34563.	34661.	34760.
12170.					34000.	34098.	34197.	34295.	34394.	34493.	34591.	34690.	34788.
12180.					34028.	34126.	34225.	34324.	34422.	34521.	34620.	34718.	34817.
12190.					34056.	34154.	34253.	34352.	34450.	34549.	34648.	34747.	34845.
12200.					34084.	34182.	34281.	34380.	34479.	34578.	34676.	34775.	34874.
12210.					34112.	34210.	34309.	34408.	34507.	34606.	34705.	34804.	34903.
12220.					34139.	34238.	34337.	34436.	34535.	34634.	34733.	34832.	34931.
12230.					34167.	34266.	34365.	34465.	34564.	34663.	34762.	34861.	34960.
12240.					34195.	34294.	34394.	34493.	34592.	34691.	34790.	34889.	34988.
12250.					34223.	34322.	34422.	34521.	34620.	34719.	34818.	34918.	35017.
12260.					34251.	34350.	34450.	34549.	34648.	34748.	34847.	34946.	35045.
12270.					34279.	34378.	34478.	34577.	34677.	34776.	34875.	34975.	35074.
12280.					34307.	34407.	34506.	34605.	34705.	34804.	34904.	35003.	35103.
12290.					34335.	34435.	34534.	34634.	34733.	34833.	34932.	35032.	35131.
12300.					34363.	34463.	34562.	34662.	34761.	34861.	34961.	35060.	35160.
12310.					34391.	34491.	34590.	34690.	34790.	34889.	34989.	35089.	35188.
12320.					34419.	34519.	34618.	34718.	34818.	34918.	35017.	35117.	35217.
12330.					34447.	34547.	34646.	34746.	34846.	34946.	35046.	35146.	35246.
12340.					34475.	34575.	34675.	34774.	34874.	34974.	35074.	35174.	35274.
12350.					34503.	34603.	34703.	34803.	34903.	35003.	35103.	35203.	35303.
12360.					34531.	34631.	34731.	34831.	34931.	35031.	35131.	35231.	35331.
12370.					34559.	34659.	34759.	34859.	34959.	35059.	35160.	35260.	35360.
12380.					34586.	34687.	34787.	34887.	34987.	35088.	35188.	35288.	35388.
12390.					34614.	34715.	34815.	34915.	35016.	35116.	35216.	35317.	35417.
12400.					34642.	34743.	34843.	34944.	35044.	35144.	35245.	35345.	35446.
12410.					34670.	34771.	34871.	34972.	35072.	35173.	35273.	35374.	35474.
12420.					34698.	34799.	34899.	35000.	35101.	35201.	35302.	35402.	35503.
12430.					34726.	34827.	34927.	35028.	35129.	35229.	35330.	35431.	35531.
12440.					34754.	34855.	34956.	35056.	35157.	35258.	35359.	35459.	35560.
12450.					34782.	34883.	34984.	35084.	35185.	35286.	35387.	35488.	35589.
12460.					34810.	34911.	35012.	35113.	35214.	35314.	35415.	35516.	35617.
12470.					34838.	34939.	35040.	35141.	35242.	35343.	35444.	35545.	35646.
12480.					34866.	34967.	35068.	35169.	35270.	35371.	35472.	35573.	35674.
12490.					34894.	34995.	35096.	35197.	35298.	35399.	35501.	35602.	35703.
12500.					34922.	35023.	35124.	35225.	35327.	35428.	35529.	35630.	35732.
12510.					34950.	35051.	35152.	35254.	35355.	35456.	35557.	35659.	35760.
12520.					34978.	35079.	35180.	35282.	35383.	35485.	35586.	35687.	35789.
12530.					35006.	35107.	35208.	35310.	35411.	35513.	35614.	35716.	35817.
12540.					35033.	35135.	35237.	35338.	35440.	35541.	35643.	35744.	35846.
12550.					35061.	35163.	35265.	35366.	35468.	35570.	35671.	35773.	35874.
12560.					35089.	35191.	35293.	35394.	35496.	35598.	35700.	35801.	35903.
12570.					35117.	35219.	35321.	35423.	35524.	35626.	35728.	35830.	35932.
12580.					35145.	35247.	35349.	35451.	35553.	35655.	35756.	35858.	35960.
12590.					35173.	35275.	35377.	35479.	35581.	35683.	35785.	35887.	35989.

GROSS WEIGHT POUNDS	18% MAC 276.1	19% MAC 276.9	20% MAC 277.8	21% MAC 278.6	22% MAC 279.4	23% MAC 280.2	24% MAC 281.0	25% MAC 281.8	26% MAC 282.6	27% MAC 283.4	28% MAC 284.2	29% MAC 285.0	30% MAC 285.8
								MOMENT/100					
12600.					35201.	35303.	35405.	35507.	35609.	35711.	35813.	35915.	36017.
12610.					35229.	35331.	35433.	35535.	35637.	35740.	35842.	35944.	36046.
12620.					35257.	35359.	35461.	35564.	35666.	35768.	35870.	35972.	36075.
12630.					35285.	35387.	35489.	35592.	35694.	35796.	35899.	36001.	36103.
12640.					35313.	35415.	35518.	35620.	35722.	35825.	35927.	36029.	36132.
12650.					35341.	35443.	35546.	35648.	35751.	35853.	35955.	36058.	36160.
12660.					35369.	35471.	35574.	35676.	35779.	35881.	35984.	36086.	36189.
12670.					35397.	35499.	35602.	35704.	35807.	35910.	36012.	36115.	36217.
12680.					35425.	35527.	35630.	35733.	35835.	35938.	36041.	36143.	36246.
12690.					35453.	35555.	35658.	35761.	35864.	35966.	36069.	36172.	36275.
12700.	MAXIMUM TAXI WEIGHT				35480.	35583.	35686.	35789.	35892.	35995.	36098.	36200.	36303.
12710.	MODEL 551 - 12,700 LB.					35611.	35714.	35817.	35920.	36023.	36126.	36229.	36332.
12720.						35639.	35742.	35845.	35948.	36051.	36154.	36257.	36360.
12730.						35667.	35770.	35874.	35977.	36080.	36183.	36286.	36389.
12740.						35695.	35799.	35902.	36005.	36108.	36211.	36314.	36418.
12750.						35723.	35827.	35930.	36033.	36136.	36240.	36343.	36446.
12760.						35751.	35855.	35958.	36061.	36165.	36268.	36371.	36475.
12770.						35779.	35883.	35986.	36090.	36193.	36296.	36400.	36503.
12780.						35807.	35911.	36014.	36118.	36221.	36325.	36428.	36532.
12790.						35835.	35939.	36043.	36146.	36250.	36353.	36457.	36560.
12800.						35863.	35967.	36071.	36174.	36278.	36382.	36485.	36589.
12810.						35891.	35995.	36099.	36203.	36306.	36410.	36514.	36618.
12820.						35920.	36023.	36127.	36231.	36335.	36439.	36542.	36646.
12830.						35948.	36051.	36155.	36259.	36363.	36467.	36571.	36675.
12840.						35976.	36080.	36184.	36287.	36391.	36495.	36599.	36703.
12850.						36004.	36108.	36212.	36316.	36420.	36524.	36628.	36732.
12860.						36032.	36136.	36240.	36344.	36448.	36552.	36656.	36761.
12870.						36060.	36164.	36268.	36372.	36476.	36581.	36685.	36789.
12880.						36088.	36192.	36296.	36401.	36505.	36609.	36713.	36818.
12890.						36116.	36220.	36324.	36429.	36533.	36638.	36742.	36846.
12900.						36144.	36248.	36353.	36457.	36562.	36666.	36770.	36875.
12910.						36172.	36276.	36381.	36485.	36590.	36694.	36799.	36903.
12920.						36200.	36304.	36409.	36514.	36618.	36723.	36827.	36932.
12930.						36228.	36332.	36437.	36542.	36647.	36751.	36856.	36961.
12940.						36256.	36361.	36465.	36570.	36675.	36780.	36884.	36989.
12950.						36284.	36389.	36493.	36598.	36703.	36808.	36913.	37018.
12960.						36312.	36417.	36522.	36627.	36732.	36837.	36941.	37046.
12970.						36340.	36445.	36550.	36655.	36760.	36865.	36970.	37075.
12980.						36368.	36473.	36578.	36683.	36788.	36893.	36998.	37104.
12990.						36396.	36501.	36606.	36711.	36817.	36922.	37027.	37132.
13000.						36424.	36529.	36634.	36740.	36845.	36950.	37055.	37161.
13010.						36452.	36557.	36663.	36768.	36873.	36979.	37084.	37189.
13020.						36480.	36585.	36691.	36796.	36902.	37007.	37112.	37218.
13030.						36508.	36613.	36719.	36824.	36930.	37035.	37141.	37247.
13040.						36536.	36642.	36747.	36853.	36958.	37064.	37170.	37275.
13050.						36564.	36670.	36775.	36881.	36987.	37092.	37198.	37304.
13060.						36592.	36698.	36803.	36909.	37015.	37121.	37227.	37332.
13070.						36620.	36726.	36832.	36937.	37043.	37149.	37255.	37361.
13080.						36648.	36754.	36860.	36966.	37072.	37178.	37284.	37389.
13090.						36676.	36782.	36888.	36994.	37100.	37206.	37312.	37418.
13100.						36704.	36810.	36916.	37022.	37128.	37234.	37341.	37447.
13110.						36732.	36838.	36944.	37051.	37157.	37263.	37369.	37475.
13120.						36760.	36866.	36973.	37079.	37185.	37291.	37398.	37504.
13130.						36788.	36894.	37001.	37107.	37213.	37320.	37426.	37532.
13140.						36816.	36923.	37029.	37135.	37242.	37348.	37455.	37561.
13150.						36844.	36951.	37057.	37164.	37270.	37377.	37483.	37590.
13160.						36872.	36979.	37085.	37192.	37298.	37405.	37512.	37618.
13170.						36900.	37007.	37113.	37220.	37327.	37433.	37540.	37647.
13180.						36928.	37035.	37142.	37248.	37355.	37462.	37569.	37675.
13190.						36956.	37063.	37170.	37277.	37383.	37490.	37597.	37704.
13200.						36984.	37091.	37198.	37305.	37412.	37519.	37626.	37732.
13210.						37012.	37119.	37226.	37333.	37440.	37547.	37654.	37761.
13220.						37040.	37147.	37254.	37361.	37468.	37576.	37683.	37790.
13230.						37068.	37175.	37283.	37390.	37497.	37604.	37711.	37818.
13240.						37096.	37203.	37311.	37418.	37525.	37632.	37740.	37847.
13250.						37124.	37232.	37339.	37446.	37553.	37661.	37768.	37875.
13260.						37152.	37260.	37367.	37474.	37582.	37689.	37797.	37904.
13270.						37180.	37288.	37395.	37503.	37610.	37718.	37825.	37933.
13280.						37208.	37316.	37423.	37531.	37639.	37746.	37854.	37961.
13290.						37236.	37344.	37452.	37559.	37667.	37774.	37882.	37990.

GROSS WEIGHT POUNDS	18% MAC 276.1	19% MAC 276.9	20% MAC 277.8	21% MAC 278.6	22% MAC 279.4	23% MAC 280.2	24% MAC 281.0	25% MAC 281.8	26% MAC 282.6	27% MAC 283.4	28% MAC 284.2	29% MAC 285.0	30% MAC 285.8
								MOMENT/100					
13300.						37264.	37372.	37480.	37588.	37695.	37803.	37911.	38018.
13310.						37292.	37400.	37508.	37616.	37724.	37831.	37939.	38047.
13320.						37320.	37428.	37536.	37644.	37752.	37860.	37968.	38075.
13330.						37348.	37456.	37564.	37672.	37780.	37888.	37996.	38104.
13340.						37376.	37484.	37593.	37701.	37809.	37917.	38025.	38133.
13350.						37404.	37513.	37621.	37729.	37837.	37945.	38053.	38161.
13360.						37432.	37541.	37649.	37757.	37865.	37973.	38082.	38190.
13370.						37461.	37569.	37677.	37785.	37894.	38002.	38110.	38218.
13380.						37489.	37597.	37705.	37814.	37922.	38030.	38139.	38247.
13390.						37517.	37625.	37733.	37842.	37950.	38059.	38167.	38276.
13400.						37545.	37653.	37762.	37870.	37979.	38087.	38196.	38304.
13410.						37573.	37681.	37790.	37898.	38007.	38116.	38224.	38333.
13420.						37601.	37709.	37818.	37927.	38035.	38144.	38253.	38361.
13430.						37629.	37737.	37846.	37955.	38064.	38172.	38281.	38390.
13440.						37657.	37765.	37874.	37983.	38092.	38201.	38310.	38419.
13450.						37685.	37794.	37903.	38011.	38120.	38229.	38338.	38447.
13460.						37713.	37822.	37931.	38040.	38149.	38258.	38367.	38476.
13470.						37741.	37850.	37959.	38068.	38177.	38286.	38395.	38504.
13480.						37769.	37878.	37987.	38096.	38205.	38315.	38424.	38533.
13490.						37797.	37906.	38015.	38124.	38234.	38343.	38452.	38561.
13500.		MAXIMUM TAXI WEIGHT				37825.	37934.	38043.	38153.	38262.	38371.	38481.	38590.
		MODEL 550 - 13,500 LB.											

ANSWERS TO STUDY QUESTIONS

CHAPTER I

1.	False	11.	True
2.	False	12.	False
3.	False	13.	True
4.	True	14.	False
5.	False	15.	False
6.	False	16.	False
7.	True	17.	True
8.	False	18.	False
9.	True	19.	False
10.	True	20.	True

CHAPTER III

1.	C	9.	C
2.	D	10.	A
3.	B	11.	A
4.	C	12.	B
5.	B	13.	B
6.	A	14.	B
7.	B	15.	A
8.	A		

CHAPTER II

1.	B	11.	C
2.	A	12.	A
3.	C	13.	A
4.	C	14.	B
5.	A	15.	C
6.	A	16.	C
7.	B	17.	B
8.	A	18.	B
9.	A	19.	C
10.	B	20.	A

CHAPTER IV

1.	B	11.	B
2.	C	12.	C
3.	A	13.	B
4.	C	14.	A
5.	A	15.	B
6.	B	16.	B
7.	B	17.	B
8.	B	18.	B
9.	B	19.	A
10.	D	20.	B

Aircraft Weight and Balance
ANSWERS TO FINAL EXAM

1. B
2. C
3. B
4. B
5. B
6. A
7. A
8. B
9. C
10. B
11. C
12. A
13. A
14. A
15. B
16. A
17. C
18. C
19. C
20. C